# Safety
## is a People
# Business

T0198085

## Michael V. Manning, Ph.D.

**GOVERNMENT INSTITUTES**
**ROCKVILLE, MD**

Government Institutes, Inc., 4 Research Place, Rockville, Maryland 20850, USA.

*Library of Congress Cataloging-in-Publication Data*

Manning, Michael V., 1943-
        Safety is a people business: a practical guide to the human side of
safety / by Michael V. Manning.
                    p.        cm.
        Includes bibliogaphical references and index.
        ISBN: 0-86587-597-9
        1. Industrial  safety--Management.  I. Title.
    T55.M3515      1997
    658.4'08--dc21
                                            97-37710
                                            CIP

Printed in the United States of America

*To The Reader*: The people part of safety begins with you.

*To Kevin Wilt*: The most conscientious and caring Safety Director I ever trained. Simply the best!

*To Lynn Manning*: My business partner and best friend, who spent countless hours editing and revising the book, and who patiently explained to me over and over again, "Just because you understand it, doesn't mean the reader will." Her help and support were immeasurable.

Michael V. Manning, Ph.D.

# Table of Contents

# Chapter 6

**"Woe Is Me!" The Safety Director on the Safety Director ........... 79**

# Chapter 7

**"Let's Not Get Carried Away With This Stuff." The Safety Director and the Disciplinary Process ........... 93**

# Chapter 8

**"They Won't Even Come for Free Doughnuts." The Safety Director and the Safety Committee ........... 103**

# Chapter 9
**"But He's the Boss's Son." The Safety Director and Company Politics.... 113**

# Chapter 10
**"They're All Thieves and Liars." The Safety Director and Workers'
Compensation ...................................................................... 127**

# Chapter 11
**"You Just Can't Train These People." The Safety Director and Employee
Training ............................................................................. 147**

# Chapter 12
## "It's Never Their Fault." The Safety Director and Accident Investigation

# Chapter 13
## "If Only They Would Listen to Me." The Safety Director and Communication

## Chapter 14
**"It's Not That Big of a Deal." The Safety Director and Personal Commitment**

## Chapter 15
**"Now, What Is It I'm Supposed To Do Again?" Putting It All Together... 199**

# Preface

The first sentence in my book *"So You're the Safety Director!"* reads, "This book is not intended for the safety professional." Well, *Safety Is A People Business* is intended not only for the safety professional, but for anyone who is involved in the protection of the lives and health of employees, regardless of their level of knowledge and expertise. It is written for those who may one day hear the words, "Oh, by the way, you're now responsible for safety," as well as those who hold certifications or degrees in the field of safety management.

This book was written at the request of participants in seminars based on *"So You're the Safety Director!"* The chapter titles are based on questions quoted from class participants. If all the comments could be rolled into one statement, it would be "Please write a book about dealing with people in our profession. You keep telling us that safety is a people business, so write a book about it." Here it is!

This book is aptly named. *Safety really is a people business*, but many of us responsible for safety fail to realize it. I am not opposed to colleagues who feel the need to limit our profession to an engineering discipline, but they need to understand that behavioral science plays an equally important role in what we do.

Wherever I go, whoever the client or audience is, I deal with people issues that prevent the successful protection of workers' lives and health. Why is the protection of human beings so difficult for individuals and groups to accomplish? Maybe it is because we often fail to factor in the human element. This is sad. Shouldn't people be the reason why we're doing what we're doing in the first place?

*Safety Is A People Business* should give you some insight into the human side of the safety business. What I present and discuss in these pages comes from more than 29 years of experience dealing with people in the safety profession. I hope it will help you to make your workplace safer.

Michael V. Manning, Ph.D.

# Safety is a People Business

# "My Boss Just Doesn't Care!"

# You and Your Boss

Let's begin at the beginning. OSHA stated in its health and safety guidance document in 1992 that two of the minimum requirements for an effective safety and health program are management commitment and employee involvement.

## Management Commitment

Commitment to safety is an excellent principle indeed, and the minimum a Safety Director must have in order to provide a successful workplace safety program. Many programs fail simply because they receive little or no management commitment.

Let's be sure we are using definable terms here:

> **com•mit•ment** *n.*   something one is bound to do or forbear.

Why doesn't the boss really commit to the program the way you want him or her to?

The first thing we need to look at is *why* did you become the Safety Director in the first place? The vast majority of Safety Directors are appointed in-house with little or no training or experience. The boss, usually due to a recent OSHA inspection or lack of workers' compensation coverage, determines that something needs to be done and that you are the one to do it!

The offer, if there is an offer, usually sounds like this, "Say, when you have the time, will you take care of that safety stuff?" It sounds so simple, so benign—why wouldn't you want to help out? Usually the first shock occurs when you attend a seminar to learn about the "safety stuff." I cannot begin to count the number of times participants in our seminars, particularly those taking the "OSHA Mandated Recordkeeping and Training Seminar," develop that "deer-in-the-headlights" look when it dawns on them that this is not small stuff but a full-time profession that requires immense time and knowledge.

Can an in-house employee with no knowledge or training ever hope to perform the Safety Director's role? Absolutely! That person will need training, consulting, and mentoring, but, yes, he or she can perform as well as those with a degree in occupational safety.

But the first thing you need to know is what you don't know. And just as importantly, what your boss doesn't know. Safety is a profession, not because those of us involved in it want recognition from society as a whole and from one another individually, but because a safety professional really needs to know "a lot of stuff." Worker safety depends on the recommendations you make. Safety is not merely common sense— yours or anyone else's. As someone once put it so succinctly, "Common sense is not that common."

## Your Safety Program and Money

Let's agree that you had no idea it was ever going to be this much work, and neither did your boss. So, the first thing you need to do is educate your boss on just what it is that you do.

All safety programs really begin with *money*. It is a sad commentary that the protection of people is rooted in money, but that is a fact we need to face. Why did your boss pick *you* to run the safety program? There is a high probability that the decision was based on the financial bottom line. Employee injuries mean high workers' compensation costs. OSHA citations are expensive and, once cited, the company believes it will always be watched by OSHA. I can count on one hand the number of companies with no workers' compensation problems and no OSHA citations that call us because they just wanted to create and maintain a safe environment for their employees. As the Safety Director of your company, you do not really need to *care* what the motivation was, as long as you *know* what it was. As I said, the most common motivation is money, so let's focus on that.

## Insurance Costs

Your boss probably has an idea from the company's insurance agent that rates are going up and something needs to be done about employee injuries. One of the first things you need to do is ask your boss why there is an increased emphasis on safety. Note: I did not say to ask, "Why the *sudden* interest in safety?" To this question your boss will likely reply that safety has always been a top priority. Don't set yourself up! Asking why there is an increase in the *emphasis* on safety implies that you already know the boss cares about employee safety. But ask about his or her motivation so that you can have a focus area—e.g., if it is workers' compensation problems, you then want to look at return-to-work programs, and/or occupational medicine. In any case, find out what the real motivation is. If, like some companies I have dealt with, your company refuses to share any information with their Safety Director, I can then assure you the motivation is money.

So we start with money! Accidents and injuries can cost your company a great deal of money.

Direct costs include:

- Damage to machinery, tools, and materials
- Insurance premiums
- Medical payments
- Workers' compensation
- Fines from regulating agencies such as OSHA

Indirect costs include:

- Profit loss due to lost production
- Post accident loss of time
- Overtime work
- Re-training expenses
- Loss of efficiency

Remember that every time someone is injured, you should picture a taxi meter running, clicking off the expense to your company. Your job, whether or not it is defined as such in a job description, is to keep management informed of "what is going on." In most instances, your boss hasn't a clue as to what a Safety Director is, let alone what one is suppose to do. Neither does he or she have the slightest idea as to what a safety program is or what it's supposed to do.

In order to let your boss know what is going on, your first step is to meet with your insurance agent or company controller to determine the following:

1. What type of workers' compensation plan you are on.

2. What your insurance coverage is costing you.

3. What your experience modification factor is. In other words, what are you paying in comparison to other, similar companies?

If possible, sit down with the agent and/or company controller and ask for an explanation of this system. Learn as much about insurance as possible so you can find out where the money is going. When I tell our seminar participants this, they roll their eyes and start grumbling. Let's establish a hard fact here in the first chapter of the book: If it were easy, everyone would be doing it! If you are to do your job adequately, you will have to learn the basics of workers' compensation insurance (See Chapter 10).

## OSHA Penalties

Enough about insurance costs—how about OSHA penalties? What do they cost? How can penalties affect your company if you are cited? Do you know what mandated recordkeeping and training you are required to perform?

Call the nearest OSHA office to determine if you are on a state plan or covered by federal OSHA. Get a copy of the OSHA Standards that apply to your type of business and learn how to read them. You'll learn how to read a CFR in the chapter on technical knowledge.

Long before you sit down with your boss to determine what you need to have in order to do your job, you must first do the aforementioned homework. Learn the cost of workers' compensation, and how your state's workers' compensation system works. Learn your state's OSHA plan and what generates an OSHA inspection. Call the state offices to gather this information.

In the initial stage, you will need to attend some seminars on what it is you are supposed to be doing. "I'm Responsible for Safety! Now What?" is a seminar based on the book "*So You're the Safety Director!*," which helps new directors learn their new role. A second seminar is called "Mandated OSHA Recordkeeping and Training." Here is where you learn what the government expects you and your company to do.

If in doubt about where to start, simply call the local or state chapter of the National Safety Council and say, "I am the new Safety Director at my company and I need answers to some questions to help me get started." They will probably make helpful suggestions.

Okay, you have the basis of what you need to know. Now let's focus on your boss and you. Remember, he or she is a person, too.

## Four Categories of Bosses

Let's be realistic about your boss. You need to determine what he or she does or doesn't know. Your boss will fall into one of four categories:

1. Your boss isn't informed, but cares.

2. Your boss isn't informed and doesn't care.

3. Your boss is informed and cares.

4. Your boss is informed and doesn't care.

Let's start with the first, *"Your boss isn't informed, but cares!"* How are you going to inform this type of boss of the role you perform, the need for the program, and how much time it will take?

First, calculate the cost of insurance covering employees injuries. What are your workers' compensation premiums costing you? Look at your experience modification factor comparing your costs of injuries to that of a similar company. Look at your insurance reserves. How high are they and why are they that high? Again, all of this information can be obtained from your insurance agent or company controller. Money talks, so talk money.

Next, compute your company's incident rate for recordable injuries, lost time injuries, and work days lost. Compare your Standard Industrial Classification (SIC) code to other like businesses to make a comparison. I recommend my book, *"So You're The Safety Director!"* if you need help with incident rates and other initial stage preparation.

Now is the time to let your boss know what OSHA expects from your company. Obtain information on what OSHA recordkeeping and training requirements are and determine if your company is in compliance. How about the thousands of OSHA standards that have to be met? If you are new to safety, I strongly recommend that you contact your local safety council or safety engineering firm to do a Risk Assessment/Mock OSHA inspection of your facility. I very seldom, if ever, visit a site or take on a client without a Risk Assessment/Mock OSHA inspection. This functions as the diagnostic test of your program. What do you have? What don't you have? What are the priorities you need to establish? Contact employer's councils, safety councils, or any business organization that does this type of work. Without this diagnostic tool, it's very difficult to determine where you stand on OSHA compliance.

**Please Note:** Asking your boss for permission to bring in an outside source to determine where you stand is also a great litmus test for your boss's commitment. Let's pretend that your boss has demonstrated his comitment by okaying the risk assessment. With that information, you can begin to put together a plan to present your position in the safety program.

# Memorandum

**Date:**

**To:**     Your Boss

**From:**   You

**Subject:** Safety

Since accepting the position of Safety Director/Manager/ Coordinator for our company, I have begun research to determine where we stand on the protection of people and property. This report covers the three main areas of our company's financial exposure: Workers' Compensation Costs, OSHA Compliance, and Incident Rates comparing us to other similar businesses.

## WORKERS' COMPENSATION

After meeting with our workers' compensation claims representative and our insurance agent, the following facts are known:

- Our annual workers' compensation premium is $_____. This should be compared to our cost of $_____ from two years ago. This increase of _____ % does not correspond with our employee increase of only _____ %.

- The individual cost of workers' compensation for our _____ employees is $_____ per year. Although this does not seem that expensive on an individual basis, we can reduce this amount significantly.

- Our workers' compensation experience modification factor, (a figure used to measure us against other types of business with the same type of exposure) is _____. This simply means that for every $1.00 we spend on workers' compensation, we must pay an extra $_____ because of our poor injury rate.

- The average employee injury cost for our company is $_____.

- The average workers' compensation insurance reserve is $_____.

## OSHA COMPLIANCE

The results of our Risk Assessment/Mock OSHA inspection were startling. The following is a recap of the lengthy report furnished by the safety engineering firm conducting the mock OSHA inspection:

- An estimated OSHA penalty of $_____ would be assessed against our company as a result of noncompliance.

- We are not in compliance with _____ different training standards. The estimated penalty is $_____.

- We are not in compliance with _____ recordkeeping standards. The estimated citations are $_____.

- The following OSHA penalties have been prioritized as "must do" and the citations are as follows:

1. No written Hazard Communication Program. $_____

2. Seven different unguarded machines. $_____

3. Two blocked exits in the plant. $_____

4. Improper storage of flammable liquids. $_____

5. Six specific citations on unsafe walking/working surfaces.
   $_____

These are only the top five priorities of the total of _____ citations generated from the recent assessment.

**INCIDENT RATES**

By using our total number of employee hours worked for one year and a multiplier of 200 representing 100 employees working 50 weeks of the year, I am able to determine our incident rate compared with similar companies.

• Last year, _____ of every 100 of our employees were injured, compared to the national average of_____.

• Last year, of all employees that were injured, _____suffered a lost time injury, compared to the national average of _____.

• Last year, for every lost time injury our employees suffered, they were out of work _____ days, compared to the national average of _____.

**SUMMARY**

As you can see, our company is behind other comparable companies in all three areas. I would like to meet with you to further explain this information and how it was determined. Also, I would like at that time to review my proposal with you for needed corrective action.

## Examples on How to Get Support

The first thing you want to do is prepare a written report for your boss to read. You want to "prime the pump" so that you can gain more acceptance and support of the program later. In this sample memo, you can fill in the numbers that are appropriate for your company.

Please notice how I deliberately did not explain in the memo what an insurance reserve or incident rate was—those are items you want to explain in person. You want the boss to know that you have done your homework and you know what you are talking about.

What other information do you want the boss to know? There are many safety newsletters that come out bi-monthly, monthly, or weekly. You need to start receiving at least two of them. Read them and then make a copy for your boss with specific sections highlighted that you want him or her to see. Your boss is very busy and you have a limited time to get your point across.

If your boss isn't informed but cares, this material will give him or her the information he or she needs to assist your efforts. His or her commitment to the program will assist you immeasurably.

So far we have focused on the boss who isn't informed, but cares! If your boss falls into that category then you have a good chance of convincing him or her of the need for a commitment to safety.

But how do you know he or she falls into that category? To begin with, it will be very difficult to determine whether your boss does or doesn't care until you attempt to inform him or her. I've worked with many new Safety Directors who believe their bosses didn't care until they did a good job of informing them. Conversely, I have found many new Safety Directors who really thought they had all the support in the world when there was none. So let's operate under the premise that the boss cares until he or she proves otherwise.

If your boss now has the safety cost information and is asking for options, this is the time to prove that you can realistically do the job with the financial support offered. Please understand that I don't mean for you to attempt to protect lives and property with little or no money. On the other hand, many Safety Directors feel that if they don't get the whole pie, a slice or two isn't worth taking. If your boss is giving you signals that he or she wants to correct a problem, then go along with the financial restraints, at least for a while. Remember, you have just understood the need for the safety program and, hence, you are a convert. Your boss, however, may need a little more time to assimilate this information. Don't overwhelm him or her. Give it time.

Where do you focus your time, energy, and money? Unless there is a specific imminent danger situation requiring your immediate attention, your focus should be on training. Train, train, and train some more! Stop and think for a moment—where did your employees, supervisors, and executive staff learn to work safely? If they weren't in the military, or didn't take a shop class in high school, they likely weren't taught safe work practices. The beauty of OSHA's training requirements is that they mandate that you teach your employees the hazards associated with their jobs. This initiates a learning cycle that is beneficial for everyone involved. Again, call your safety council or safety engineering firm to make arrangements for the education of your employees. Please spend some time on this book's chapter on safety training.

## Five Questions You Need to Ask Yourself

So far I have focused on the boss who *"Isn't informed, but cares!"* Now, what do you do with the boss who *is informed and doesn't care* or the one who *isn't informed and doesn't care?* Ask yourself these five questions:

1. *Why were you selected as the Safety Director?*

2. *What authority and responsibilities were you given?*

3. *What support were you promised?*

4. *What do you want to accomplish?*

5. *What does your boss really want from the Safety Program?*

Please take a few minutes and think through those questions.

Let's look at the five questions more closely.

1. *Why were you selected as the Safety Director?* Come on now, don't be coy. You know why! It is because of your experience, your knowledge, your ability to respond to requests, or your knowledge of a similar task such as personnel or workers' compensation administrator. What if you are just uncomfortable saying "no" and your boss knows that? Whenever I ask that question in a seminar, I get about five minutes of tap dancing, and then the truth comes out. Now compare the reason you came up with to the one your boss told you. Usually your reason is more honest than what you will hear from your boss. Would your boss tell you the real reason you were selected?

2. *What authority and responsibilities were you given?* I bet you didn't ask your boss about those things when you took the job. Why would you? How were you to know the amount of work involved? Now that you know to ask, are you going to do it? Are you going to establish some boundaries in your new role or just go along and see how things work out? If you don't establish your authority, you will constantly be bumping into other people who are going to tell you the ever-popular phrase, "You're not my boss." You will then be camped out in your boss's office, either telling on other people or asking for permission to tell other people what to do. This will make you appear to be weak and

ineffectual. Now is the time to find out if your boss cares by determining how much authority you have.

After you determine that, find out how much responsibility you have. Be careful on this one. If you don't have the needed authority to carry out your safety program, then how can you be held responsible for it? Easy—it's done all the time! The number one complaint and fear of Safety Directors, particularly new Safety Directors, is that when a catastrophe occurs, they will be left holding the bag! "My boss will claim that he or she never knew these unsafe conditions existed and these unsafe acts were committed." "I'll be the fall guy and there is nothing I can do about it." Yes, there is! But first you must determine if you have total responsibility. Listen to what is being told to you, not only with your ears but with your intellect. Is it a gray, mushy explanation of how you won't need that much authority and your only responsibility is for what you can fix? Can you really work and succeed in that environment? Will your boss give you the authority and responsibility you need? There will be more discussion later on being the "fall guy."

3. *What support were you promised?* Again, did you ask for top management support or just assume that you would get it as the job went along? How about now? Are you prepared to say "I need your personal support in these safety matters"? What kind of support do you feel you need? Immediate access to your boss, financial, technical support—what is it you need to perform your job and are you prepared to ask for it? Is your boss prepared to give you the support you need?

4. *What do you want to accomplish?* When the job was given to you, what did you feel you wanted to accomplish? Most of our participants state that they never really considered that question, but rather they looked at the role of Safety Director as a task rather than a profession. Like filing papers or painting a barn, "It was something the boss wanted me to do, so I just said okay and started doing it." Well, now

the question is being asked, "What do you want to accomplish and can it be accomplished with the support you have?"

5. *What does the boss really want from the Safety Program?* When you were selected for this position, was what was expected of you explained? Did you ask? Was there a discussion of specifics or just generalities? What do you need to do to please your boss in your new role? Do you know? Does he or she know? Is the Safety Program just some obscure thing floating around, that the company has and that you're now responsible for? Do you and your boss agree on what is needed? Is it a program to save money only? What will your boss's response be to, "When will you consider the Safety Program successful?" Before you ask the question, determine the answer yourself.

This job is getting a lot more complicated than you thought, isn't it? Here you were, just doing your previous job and somebody gave you another hat to put on—called safety—and now you are starting to doubt, or at least question, your boss's motives and commitment. Well, now is the time—you and you alone must decide if you are going to be able to fulfill the role with the commitment you receive from your boss.

Trial and error methods are fine, but eventually you may have to say to yourself, "I just can't attempt to protect these people with one hand tied behind my back!" When I discuss this issue with seminar participants, they get very quiet. It's as if I have trampled over sacred burial grounds.

Most Safety Directors don't want to address these issues. They have no problem complaining about how difficult it all is, but addressing them is a whole other matter.

Let's look at the problem head on. Either you asked for, volunteered for, or were given the job of Safety Director. Okay, now you have it. What are you going to do with it? In this profession, you must be very honest with yourself. Can you protect people and property adequately with the support system and commitment that you have? It's irrelevant whether it would work for someone else or not, how about you? Is there enough commitment and support for you to feel comfortable doing the job? Be honest with yourself, and what is more important, be honest with those employees who are depending on you to protect them from workplace injuries and death. We aren't selling mutual funds here, we're talking about the lives of people. Many are your friends and co-workers. The law is very clear: you will provide them a workplace free of recognized hazards. That responsibility has shifted from your company President to you. Every 99 minutes, a worker is killed on the job; every 9 seconds, one is injured.

Let's get back to the "fall guy" syndrome. Again, it always comes up in seminars and with almost everyone I mentor. What happens when it all goes sour and people are hurt? OSHA penalties are levied and my boss pretends ignorance. "What should I do?" While you are thinking about a response to that question, let me give you the standard reply from our seminars participants. "Just cover your tail with paperwork and keep them notified of everything that's wrong and needs fixing. That way they can never say you didn't warn them." How do you feel about that response? Will it work for you? How would your boss respond to a paper avalanche orchestrated to protect you and bury him or her?

I remember two separate seminars and two distinctly different participants who gave their answers regarding that approach. One boss told an employee "I do not want any more memos regarding safety problems sent to me or anyone else. None will be written, period! Our attorney feels it could cause legal problems for us. Continue to do the best you can." The second participant said the boss stated, "You are not going to hang me with your memos. I know what you are doing and it stops now! You are the Safety Director and it's your job to solve the problems with the resources we give you!"

To think you can make a bad situation better by dumping responsibility on someone who deliberately dumped it on you in the first place is ridiculous. Those types of bosses don't want to know what's wrong, and what is more important, they don't want you to tell them! You can take the avenue of a participant in our seminar who stated, "My boss gives me no support, commitment, or money to run the program. But, I have been with the company sixteen years and have seniority and a good pension. I'm not going to do anything to jeopardize that, including making waves over safety. If people get hurt, there is nothing I can do about it." What is your opinion of this attitude? I wouldn't want to be operated on by a surgeon or fly with a pilot with that approach, would you?

What do you do? You must make the decision if you are going to have the support and commitment to do the job. You can't get pears from an apple tree—they don't grow there! The time may have come to tell your boss that you can't do the job with the restrictions you have and ask that someone else be appointed.

## Take This Job, Please!

So, maybe you feel that the job of Safety Manager should be given to someone else. Participants fall into two categories when this is presented:

A. Complete relief. I look at their faces full of frustration, fear, and fatigue and it's almost as if they are transformed.

B. Complete anger that I would suggest they put their positions on the line by actually telling the boss they can't do the job with no support or commitment. Only you know what you will do after you determine where your boss stands. But, remember that the decision you make will affect not only you but all the employees depending on you. If there is a wall between what you want to accomplish and its completion, you will eventually throw up your hands in despair anyway. Think long and hard about your decision and decide what's best for you and those employees who you were going to help. You owe them that much. Remember, safety is a people business!

Let's now go the person who is really in charge—the Plant Manager.

# 2

# "The Plant Manager Is A Dictator."

# The Safety Director and the Plant Manager

## What the Plant Manager Thinks of You

Your boss has given you the support and commitment you need. You have the green light to go out into the workplace and start your program to save lives and property. You next need to move down the food chain to the person who is really in charge of all those people—the Plant Manager. Also called the Plant Superintendent or Vice President of Manufacturing, this is the individual who is responsible for all areas of plant operations. This person is accountable for getting out a quality product on time.

What do you think the biggest problem in doing a Plant Manager's job is? He or she will tell you that, "It's the people! They're late to work, if they come at all. They test positive for drugs or alcohol. They don't care about quality and most of them have no work ethic." Plant Managers spend ninety percent of their time with their ten percent problem employees. They aren't supposed to be Human Resource Directors and chaplains, but they find they are handling the people problems more and more. Their key motivation is delivering a quality product on time.

Remember how your boss told you that safety was a priority, or maybe even said it's the number one issue. What do you think your boss tells the Plant Manager? Not that! The Plant Manager is told to get the product out on time and to try to do it safely.

Plant Managers know what the role of the Safety Director is—to cause personnel problems, slow down production, second guess every decision made, and constantly shove OSHA standards in their faces that have to be followed. Other than that perception, you and the Plant Manager will get along famously. Now do you understand why it was so important to get that commitment and support from your boss? Regardless of your past association with Plant Managers, your new role will change everything. You're going to get into their rice bowl, and they aren't going to like it. What do you do?

## Approaching the Plant Manager

Start off by reaching an understanding with the Plant Manager that you represent the company's commitment to providing a safe working environment for people. You do this with a memo signed by your boss stating the ground rules. Be careful here—if it's too vague, you'll never be able to enforce and accomplish what you want. If it's too specific, you will hear nothing but a literal translation of it and will get no conceptual acceptance. How about a joint effort between you and the Plant Manager? You, your boss, and the Plant Manager should meet to discuss how the role of the Safety Director and Plant Manager can help provide your company with a safe work environment. The following sample memorandum may help.

Remember, the title of this book is *Safety Is A People Business*. Nowhere else is that philosophy better reflected than with the Plant Manager. Dealing with Plant Managers and their motivators is difficult at best, and can be crippling at worst. Remember this also—there will be no safety program if there is no production, and no job if there is no company.

**MEMORANDUM**

**Date:**

**To:**     Plant Manager

**From:**   Big Boss

**Subject:**  Safety Program

As we discussed, our company has a legal, moral, and ethical responsibility to provide our employees with a safe work environment. To better fulfill those responsibilities we have asked (your name) to become our new Safety Director. Although he (she) will not be the Safety Director full time, he (she) will have overall responsibility for the administration of the program. Since he (she) will be dealing with our employees, I feel that the three of us should meet to discuss your participation in safety activities. The following is a list for your review prior to our meeting.

**Plant Manager Responsibilities**

1. Assist the Safety Director in adoption and enforcement of company rules and regulations.

2. Oversee supervisors to insure acceptance, adoption, and enforcement of the company safety program by production and engineering of new equipment.

3. Accompany the Safety Director on all production-related safety inspections.

4. Be a permanent member of the company safety committee.

Empathy is the one quality that Safety Engineers who are just out of college seem to lack, particularly with the Plant Manager. New Safety Engineers can spout OSHA standards like androids. But they have no experience to fall back on to remind them that others have a job to do as well. Not only do unsafe acts and unsafe conditions kill employees, a lack of real world experience kills as well. Trying to bulldoze your perception of a safe work environment through all the players in the program, no matter how much support and commitment you have, will spell disaster for your efforts in the end.

Your boss, no matter how motivated, will not continue to step on the Plant Manager so you can have your way. The Plant Manager is the one person you have to get along with. If there is a dispute between the Plant Manager and you, you will probably lose. The boss watches the Plant Manager very carefully to see that a quality product gets out on time. It's going to be more difficult to replace a Plant Manager than a Safety Director. With that being said, you still have an important role to serve.

Let's now go over those points in your memo to make sure we understand how to explain what it is you are requesting from the Plant Manager.

## The Four Responsibilities of the Plant Manager for Safety

### Assist the Safety Director

The Plant Manager must assist the Safety Director in the adoption and enforcement of company safety rules and regulations.

Plant Managers need to review these safety rules with you if you are just starting to write them. Remember, *their* employees are going to have to follow them. *Their* foremen are going to have to enforce them. If the rules are already written, how are you going to demand compliance? You may need to involve the Human Resources Director in this process. Plant Managers have no problems with safety rules, as long as these rules don't slow down production. New Safety Directors often make the statement, "I don't cut deals on safety. It's either totally safe or we shut the line down."

Let's talk a moment about discernment.

---

**dis•cern•ment**   *n.*   acuteness of perception or judgment.

---

The use of discernment among Safety Directors, either new or experienced, is a rare thing indeed. One of the biggest complaints I hear from business owners and Plant Managers is that Safety Directors don't seem to know the difference between imminent danger and de minimus exposure. If the Safety Director can't set the standard of safety in a company, who can? One of the keys for your success in dealing with Plant Managers is your ability to look at a situation and make a judgment call. Will the violation of a safety rule create an imminent danger exposure or will it simply be a less than 100 percent safe situation?

Let's look at an example. The Plant Manager has come to you asking permission to temporally ignore two written safety rules pertaining to a 70-ton heavy press, due to a shortage of trained heavy press operators. The Plant Manager has already been to the President's office and the order simply must go out today!

## Scenario #1

Two months ago, you established a two-employee rule on feeding large sheets of steel into the heavy presses. Your research, along with employee complaints, demonstrated that more emphasis on task sharing was needed in this department. Therefore you implemented the two-operator rule. Two employees are required to feed raw stock, since it would be dangerous to have one employee wrestle with it alone. The job change has been a resounding success. All heavy press operators are much happier with the assistance and their complaints of muscle soreness have dropped dramatically. Now, the Plant Manager has requested that he or she be allowed to revert back to a single operator for one day so that other operators can be freed up. How are you going to respond?

## *Scenario #2*

The heavy presses have full automatic operation so steel can be fed into the machine and stamped out quickly. A light sensing device was installed so that in the event an employee's hands entered the point-of-operation (danger zone) the machine will automatically stop. Regrettably, the machine now has a tendency to jam, thus forcing the operator to shut it down frequently to clear it. The Plant Manager has requested that the light sensing devices be turned off for one day so that the operators can adjust the stock without constantly shutting down the machine. What is your answer?

In Scenario #1, will the elimination of one operator for an eight-hour period really create an unsafe condition for the employee? Will the employee be sufficiently at risk to prevent you from allowing the temporary termination?

Almost 100 percent of those individuals who have any number of years experience in the safety field will allow the elimination of the one operator for a small amount of time. Why? Two reasons, really. The safety consequences would ordinarily not harm the worker, and the Plant Manager needs your assistance. You have him or her in a tough spot; will you be adaptable enough to work with the Plant Manager? I agree that allowing the elimination of one worker will not create a serious enough exposure to harm the other. Many new Safety Directors in our seminars vehemently disagree. They claim we are setting a bad precedent. They feel there are never any exceptions to safety. Oh, if that were only true! But here in the real world, it is vital to understand the needs of Plant Managers, as long as they appreciate your needs, too.

In Scenario #2, the answer is an absolute no! The employee is exposed to an imminent danger risk, and the guarding of heavy presses is mandatory without exception, even if production time is slowed.

From this example, you can see why it is important to use discernment in your day-to-day dealings with people. Your ability to use astute perception and judgment will benefit you as much as your ability to quote OSHA standards. Discernment—work on it, practice it!

## Oversee Line Supervisors

Plant Managers must oversee Line Supervisors to ensure their acceptance, adoption, and enforcement of the company safety program.

In most cases, if not all, the production plant foremen work for the Plant Manager. These are his or her employees; he or she is their boss. You are not! The Safety Director is the company's specialist or technician on the subject of safety. Only in very rare cases is safety a line function; usually it is a staff function.

The Plant Manager's control of the foremen is very critical to your success. Foremen or supervisors can make or break a safety program. Their acceptance and enforcement of the program can spell success. Their dismissal or refusal can destroy it. The Plant Manager needs to understand this.

Plant Managers do not want to get involved in a "monkey-in-the-middle" situation where they must constantly be defending you to their Line Supervisors and vice versa. Plant Managers want a clear understanding of what the rules are for the supervisors, so they can meet with them to convince them of the need for the program and you. Understand that the Plant Manager needs these supervisors to get a quality product out on time. Experienced supervisors are very difficult to find, and it's time consuming to bring them up through the ranks. Give heed to the Plant Managers' reservations about placing more work on their already overburdened supervisors.

## The Seven Guidelines for Line Supervisors the Plant Manager Can Live With

Establish some basic guidelines for Line Supervisors to follow that the Plant Manager can live with.

Here are some guidelines that have been used successfully nationwide:

1. *Do the Line Supervisors ensure their employee's compliance with safety rules and regulations?* Again, look at these guidelines as if you are the Plant Manager. As you walk through your facility, are the Supervisors ensuring that the employees are following safe practices? Is there enforcement?

2. *Do the Line Supervisors tolerate unsafe behavior for the sake of production?* Are "high producers" in their departments permitted to cut corners in their adherence to safety rules and regulations? Are the Line Supervisors equitably enforcing the intention of the safety program?

3. *Are the Line Supervisors using their acquired safety knowledge?* Your supervisors must be trained in safety on a variety of subjects, not only for OSHA compliance, but for company wide safety application. Once this training has been given, are they using it? If not, then their employees will not be following safe practices.

4. *Do Line Supervisors listen to their employees' ideas on methods to improve safety in their department?* One of the best sources of

information on unsafe acts and unsafe conditions (the only two causes of accidents) is the line employee. Their supervisors must encourage and listen to their ideas.

5. *Do Line Supervisors solve safety problems actively, resourcefully, and independently?* Must every safety issue, no matter how trivial be brought to your attention for correction? If so, the Line Supervisors are only paying lip service to their commitment to the program.

6. *Do Line Supervisors make their rounds?* By this I mean, are they walking through their department looking for safety-related items that could cause accidents/incidents? If they normally make the rounds for production and quality, why not for safety?

7. *Do the Line Supervisors properly handle accident/incident investigations and other safety paperwork?* Safety programs create paperwork, and a real test of whether or not supervisors are involved is whether they will take the time for it.

These seven guidelines can assist you and the Plant Manager in assuring that Line Supervisors will accept, adopt, and enforce your company's Safety Program. Even more important, the Plant Manager can "sell" this as an equitable set of guidelines to supervisors.

## Safety Inspections

The Plant Manager should accompany the Safety Director on all production-related safety inspections. The vast majority of your exposure, when accidents/incidents occur, will be from the production floor. That is where you are going to be spending the majority of your inspection time. Remember this is the Plant Manager's turf. This is their playing field. One of the biggest complaints I hear from Plant Managers is that Safety Directors operate, or attempt to operate, independently of them.

An excellent example of being on the Plant Manager's turf is the safety inspection. Whether you do daily, weekly, monthly, or quarterly inspections it just makes good sense to notify and involve the Plant Managers. My suggestion is that if you use a formal monthly system, make them a part of your inspection team. If you conduct frequent "rounds," then simply let them know first that you are going out. It's as much a matter of courtesy as it is good management practice. Nothing will assist you more in the inspection process than having Plant Managers accompany you. After all, the employees work for them! Also, Plant Managers know the process, equipment, and personnel. They will know the difference between when something is legitimately wrong and when it's just sloppiness or carelessness—a difference you might not know otherwise. Remember, you want the Plant Managers with you. Don't try to bypass them. You're charged with protecting their people.

## Safety Committee

The Plant Manager should be a permanent member of the company safety committee. Let's begin by defining what a safety committee is supposed to do.

Your safety committee is a tool used to let those around you know what it is you are doing and to get input from them for the betterment of the program. The safety committee is where your safety team is built. Please spend some time reading and re-reading the chapter in this book on safety committees—it is vital to your success.

The tasks of this committee should be as follows:

1. Reviewing new safety policies.

2. Reviewing safety training needs.

3. Informing the committee of new federal and state standards.

4. Reviewing accident/incident data.

5. Identifying safety-related problems and the correction needed.

6. Reviewing unsafe acts and unsafe conditions.

7. Reviewing accident investigation reports.

8. Reviewing anything that you, the Safety Director, feel needs to be reviewed, discussed, or debated. This is your committee.

Can you imagine doing all of this without the advice, input, suggestions, and observations of the Plant Manager, who is the one person you definitely need on the committee? He or she is the person who has the authority to get things done on the plant floor. Don't attempt to bypass Plant Managers on the Safety Commitee. Everything that you do in this committee will in one way or another affect them and their employees. Attempt to use their knowledge and experience. It will help your program more than you can imagine.

## Use of Discernment with the Plant Manager

You now know how important it is to involve the Plant Manager. Your boss has sent him or her a memo listing the four priority items. But let's say that you have bent over backwards to get along and work with the Plant Manager, but it's just not succeeding. What do you do? I have had countless new Safety Directors say that getting along with the Plant Manger is like hitting a brick wall. Some Plant Managers just won't play ball, are passive/aggressive, or even openly hostile or belligerent. Or maybe they are always "too busy" and attempt to patronize you and your position. What happens when you have attempted to talk straight and play fair, but they haven't? What do you do then? The obvious answer is to sit down with your boss and have a chat.

When this solution is offered in seminars, the participants say things like, "The Plant Manager will get even with me somehow." " I can't point out specifics, but I don't want to rock the boat." The last statement always surprises me. If an employee is severely injured or killed, won't that place your job in jeopardy or cause some boat rocking? When OSHA issues a large penalty, won't that put your job in jeopardy?

There is no easy way out of this situation with the Plant Manager, except the truth. What is it that you need from the Plant Manager that you aren't getting? Make a list. The list items may include concrete examples of failure to enforce safety rules, policies, or procedures, or not attending meetings. They can also be your perception of the Production Manager's failure to support your efforts.

The biggest problem Safety Directors usually have with Plant Managers is their attitude. If their attitude is not supportive of the program, then the program will inevitably fail.

Now you have made up a list of specifics and perceptions, but you still can't bring yourself to meet with your boss to "rat on" the Plant Manager. What then? Well, you can do nothing and hope that the situation gets better. You can meet with the Plant Manager one last time and lay your cards on the table. That, however, often comes off as a threat. You can visit your boss and say the job is just more work than you can handle. Or you can sit down and talk straight. Tell your boss what the problem is, what you need, and present options for consideration. Saying nothing won't work. When an employee is killed or OSHA levies a large fine on your company for failure to comply, your boss will ask, "Why didn't you tell me?" What is your response going to be then? "I didn't want to rock the boat"? When it comes to the protection of human life, talking straight and playing fair is not an option, it is an obligation.

I know that dealing with belligerent or overbearing Plant Managers can be a nightmare. There are solutions to this problem, however, without looking the other way when employees are at risk. Really decide if you still want to be the Safety Director. Sit down with your boss and talk straight and play fair about the difficulty you are having with the Plant

Manager. If the cost is going to be too high for you, and there are situations when it just is, then resign from this position for your own good, and for the good of the employees you are charged with protecting. It really will be in your best interest and the interests of the company.

Now let's focus on the Line Supervisor, the person in charge of people.

# 3

# "They Just Smile and Nod."
# The Safety Director and First Line Supervisors

The reason you keep reading that First Line Supervisors are the key to your program is because it's true! They are the ones who can make or break the program simply with tone of voice or facial affect. Regardless of what you read about teams and their role, First Line Supervisors are the people who have daily contact, authority, and responsibility for the line employee and they are the ones who will help or harm the program.

What motivates them and makes them want to support you? The first question a Line Supervisor is going to ask when the safety program is presented is usually "What's in it for me?" or "What happens if I don't comply?" All the more reason you should start with the Plant Manager. This is the person the First Line Supervisor works for—not you! You are the technical advisor, not the boss—you can only report to the Plant Manager as to compliance or noncompliance of a Line Supervisor.

## Seven Guidelines for Line Supervisors

Let's review the seven guidelines for Line Supervisors previously listed in Chapter #2:

1. *Do the Line Supervisors ensure their employee's compliance with safety rules and regulations?* Again, look at these guidelines as if you are the Plant Manager. As you walk through the facility, are the Line Supervisors assuring that their employees are following safe practices? How is enforcement handled?

2. *Do the Line Supervisors condone unsafe behavior for the sake of production?* Are "high producers" in departments permitted to cut corners in their adherence to safety rules and regulations? Are the Line Supervisors equitably enforcing the intention of the safety program?

3. *Are the Line Supervisors using their acquired safety knowledge?* Line Supervisors must be trained in safety on a variety of subjects, not only for OSHA compliance but also for company safety application. If the Line Supervisors are not practicing safety, then their employees likely also are not.

4. *Do the Line Supervisors listen to their employees' ideas on ways to improve safety in the department?* One of the best sources of information on unsafe acts and unsafe conditions (the only two causes of accidents) is the line employee. Their Line Supervisor must encourage and listen to their ideas.

5. *Do the Line Supervisors solve safety problems actively, resourcefully, and independently?* Must every safety issue, no matter how trivial, be brought to your attention for correction? If so, the Line Supervisors are only paying lip service to their commitment to the program.

6. *Do the Line Supervisors make their rounds?* By this I mean are they walking through their department looking for safety-related items that could cause accidents/incidents? If they normally make the rounds for production and quality, why not for safety?

7. *Do the Line Supervisors properly handle accident/incident investigation and other safety paperwork?* Safety programs create paperwork and a real test of whether or not Line Supervisors are involved is if they will take the time to complete it.

Here comes the "people part" of safety. The Plant Manager has been bought into your safety program and has met with the Line Supervisors and reviewed the seven guidelines. If you were the Line Supervisor, you would have smiled and nodded your compliance to please the Plant Manager. After the meeting, you would find it interesting that for years you, as a supervisor, have been rewarded for production and quality, and now safety is being dumped on you. What you don't understand is why the Safety Director doesn't handle these things and let you do what you do best? As a Line Supervisor, you will, of course, continue to look after your employees' safety, as you always have. But your primary focus needs to be production and quality. If safety is such a hot item, you will soon see it reflected by the Plant Manager. The best course to take now is to smile and nod and see what happens. Well, Safety Director, what happens when Line Supervisors, for whatever reason, don't comply?

Getting support, commitment, and acceptance from the boss and the Plant Manager is relatively simple compared to obtaining it from Line Supervisors and line employees. When starting with the Line Supervisors, you need a new acronym for your vocabulary and a new system in the people part of your program. I call it **RACK—Responsibility, Accountability, Consequences, and Knowledge**. It is a strategy I created that has been used successfully for years. Let's define the terms that make up the acronym.

First Line Supervisors were, are, and always will be the key to a successful safety program. No one spends as much time with the employees as First Line Supervisors. No other members of management know the line employees as well. When management does not enforce the First Line Supervisors' involvement in the safety responsibility of their line employees it is being negligent. First Line Supervisors must be answerable for the safe conduct of their employees at all times. They can never be allowed to shift that inherent and proven responsibility to the Safety Director or other staff.

Most Line Supervisors want to be liked, want a minimum of personnel turbulence in their department, and want a happy, productive workforce. Safety is hard work, requiring the firm, fair, and consistent monitoring of employees. The Line Supervisor's perception is that safety concerns take time away from the production task at hand. This is perception, however, and not reality. First Line Supervisors do best what they get rewarded most for. If a good safety record is where the reward lies, then that is where their focus will be. But American business is based on the bottom line and that bottom line is the number of units produced, not the entries on the 200 log. First Line Supervisors often get a schizophrenic message from management—produce, produce, produce—and try to do it safely.

## Safety Responsibility of the Line Supervisors

To be responsible is to be the cause or source of something, to be answerable for its results. First Line Supervisors are responsible for the production, quality, and—yes—*safety* of their employees. They are responsible for every stage of the program, including the enforcement and application of the safety rules and policies. You as the Safety Director/ Manager/Coordinator are not! You do not have day-to-day, hour-by-hour contact with the employees like supervisors do!

Regardless of arguments, whining, or pleading you might hear from the Line Supervisors, they are ultimately responsible. It simply can't be any other way. Remember, *Safety Is A People Business*—the monitoring, measuring, and controlling of people. Only Line Supervisors are in the position to do that.

I can't believe the number of companies that have tried to placate their Line Supervisors by using safety teams, ad hoc committees, or task groups to supercede the ultimate responsibility of Line Supervisors. After failure, these companies always return to what works—the Line Supervisor's day-to-day contact with the employee.

## Accountability of the Line Supervisor

It seems no one wants to be held accountable anymore. It has become a fact of life today. As the twentieth century winds down and we prepare for a whole new century of safety activity, we see that almost no one is being held accountable for their actions. Open a newspaper, or watch the evening news, and we discover plausible, almost believable, excuses for eliminating personal accountability.

First Line Supervisors often use the thread-bare argument, "Do you want production or safety?—You can't have both." In this way, Line Supervisors simply mirror our society today—"You cannot hold me accountable, it was not my fault. Maybe someone else's, but not mine!"

OSHA does not have to worry about whether Line Supervisors enforce safety rules because they and the insurance carriers are looking at the results. If your losses are up, so are your insurance premiums! If you are in violation of OSHA standards, you will eventually be cited. OSHA and insurance carriers can objectively hold people accountable and get away with it. Companies are either in compliance with OSHA standards, or they are not. The Safety Program is either in place and working, or it is not.

Someone must be held accountable for line safety, and that someone needs to be the First Line Supervisor. When companies take a position that there is going to be accountability, not fault finding, then employee injuries will go down. If there is going to be accountability, then there must be consequences—every action has one!

## Consequences of Line Supervisor Actions

Consequences are the result of an action or condition. Consequences in the workplace are almost always viewed as negative. This doesn't need to be the case. A positive consequence occurs when Line Supervisors reduce their department's accident/incident rates. A solid, viable safety program yields the positive consequences of a reduction in workers' compensation premiums.

OSHA's penalties are some of the best attention-getting devices operating today. In the safety profession, shoddy or inadequate work can mean the failure of a safety program, which results in crippling injuries or death. What other area in today's workplace screams for the attention to consequences more than safety? Shouldn't Line Supervisors be rewarded for going an entire year without a lost time injury? Ask their employees! Consequences must be evaluated for safety programs to succeed.

The success of any safety program is based on knowledge of what the positive and negative consequences are for the program enacted.

## Knowledge Required of Line Supervisors

The old adage that with knowledge comes power has never been more true than it is today. When we give Line Supervisors knowledge about safety, we give them control of their employees' safety, both on and off the job.

Management cannot expect accountability or attention to consequences without providing the needed knowledge that goes with the job. Where were supervisors trained to work safely if not in the military, vocational schools, or shop classes in high school? Maybe they simply weren't trained at all. American business spends millions of dollars and employee hours on production and quality training, but then grumbles about the minimum training mandated by OSHA. Line Supervisors and their employees need to be trained to work safely. It has been proven time and again that trained workers have fewer injuries. They have the right to be given the knowledge to protect themselves.

## Applying RACK to the Guidelines for Supervisors

Since Line Supervisors are responsible for the safety of their employees, we should apply RACK to their enforcement of employee compliance with safety rules and regulations.

## Responsibility

Line Supervisors are ultimately responsible for the enforcement of safety rules and regulations. Stop and think for a moment—Why is it that you can walk through the plant and find an employee not wearing personal protective equipment, not properly following safety procedures, or involved in an unsafe working condition, yet, the Line Supervisor didn't seem to notice these same situations and may be surprised that you discovered them?

The answer is simple. Either Line Supervisors are not looking for these safety issues or they are ignoring them when they do see them. Line Supervisors want to be liked by their employees. No one wants an unhappy department. Line employees are needed for production and quality. If they are unhappy or feel that they are being hounded by their supervisors, they can make life difficult in many ways. Line supervisors know this. They also know that if the employees complain long enough and loud enough about how the Safety Director and the safety program is harming production, they will eventually get someone's attention to curb your efforts.

Stay the course on this one, Safety Director! Your approach should be firm, fair, and consistent. Line supervisors need to be told by Plant Managers that the compliance enforcement arm of safety is their responsibility. If you back off this guideline, there is no reason for you to continue on with the other six. This one is a must! Line Supervisors must ensure their employees' compliance with safety rules and regulations!

## Accountability

Line Supervisors must answer for their actions or inactions regarding compliance with safety rules and regulations. Here is a very common scenario in companies:

**Safety Director:** "X was not wearing his goggles while pouring acid into the acid tank."

**Line Supervisor:** "I can believe that. I tell him and tell him but he just doesn't listen."

**Safety Director:** "What are you going to do about it?"

**Line Supervisor:** "What do you mean what am I going to do about it? I just told you, I tell him and tell him and it doesn't do any good!"

Let's stop for a minute and look at that exchange between the Line Supervisor and the Safety Director. What is wrong with that dialogue? Have you ever found yourself in that position? The Line Supervisor is putting the responsibility and accountability back on the employee, not on himself or herself. Many Line Supervisors will attempt to do this. Their reaction to being confronted with unsafe acts and unsafe conditions created by their employees is the oldest technique in the safety engineering discipline—hand washing. "I tell them, they won't do it. What do you expect from me?"

Well, what we need to expect from Line Supervisors is accountability. They don't allow their employees to come to work drunk or on drugs, be insubordinate, be excessively tardy for work, or steal from the company. Why then should they allow employees to break company safety rules with impunity? Sometimes Line Supervisors allow this because they don't equate safety with production or quality. The concept is so strange to them that it just doesn't compute. They also don't want to accept the responsibility and accountability for their employees on another issue. They feel their plate is already full enough! How, then, do we get them to be accountable and responsible for safety measures at the line level?

## Consequences

We have been taught that for every action there is a reaction. Well, for every action there are also consequences. Usually, when we discuss consequences in the work environment, we discuss them regarding the line employee, not the Line Supervisor and almost never regarding safety compliance. Earlier, we alluded to not allowing employees to come to work late, or drunk, and not allowing insubordination or stealing. How about violating production and quality standards? Isn't safety at least as important as production and quality? Any Line Supervisor permitting unsafe acts or unsafe conditions should be reprimanded, since the consequences of such unsafe conditions will almost always result in personal injury and monetary loss to the company. It really is a matter of education.

Top management views safety in what I refer to as the "hub cap mentality." Hub caps on cars do not make them run one bit better, but they do improve the appearance and the owner's pride in the vehicle. In many ways, the same applies to safety programs for companies. Hub cap safety programs are seen as a way of placating employees, meeting legal requirements, and reducing workers' compensation payments. But, are they really an integral part of the company philosophy and mission? Usually not. Therefore, the consequences of not strictly applying safety rules, principles, and standards are not the same as those consequences that result from an attention to quality products, production time, and costs. An auto mechanic will suffer more serious consequences for not replacing the brakes than for not replacing the hub caps. Please understand this is not a defense for the lack of attention to safety consequences, but rather a reason for it! So how do you, as a Safety Director, deal with it?

To begin with, understand that if your company suffers from the hub cap mentality it is not a green light for your Line Supervisors to get a free ride. They need to be educated that they have a responsibility to the safety of their employees, are accountable for that responsibility, and must suffer the consequences for not being so.

When I discuss educating Line Supervisors, I do not mean conducting a formal seminar. That should come later with their increasing need for safety knowledge. What these supervisors need to hear first of all is the financial cost of accident/incidents to the company and how the safety program will help reduce these costs. They should also be told that the consequences of their efforts will be rewarded or reprimanded at the same level as they are for quality and production. Do not become overwhelmed when the Line Supervisors do not embrace this new concept and when they claim they still do not understand the need for their involvement. It is enough that you have explained the rules to them and that they have heard these rules.

---

# SUPERVISOR SAFETY PERFORMANCE EVALUATION

Name_____ Job Title_____Department_____
Todays Date_____  Last Evaluation Date_____

| Poor | Below Average | Average | Above Average | Outstanding |
|------|---------------|---------|---------------|-------------|
| 1 | 2 | 3 | 4 | 5 |
| Weaknesses to overcome | Meets some requirements | Performance meets overall requirements | Performance above average | Outstanding performance |

## Principal Accountabilities

Explanatory comments are essential where performance is noted to be below average or poor.

1._____Ensures compliance of subordinates' activities regarding safety rules and regulations.

2._____Actively enforces department safety procedures for employees in a fair and consistent manner.

3._____Uses acquired safety knowledge to improve safety within the department.

4._____Maintains open and receptive communication with subordinates regarding department safety issues.

5._____Is resourceful and uses initiative in resolving department safety problems.

6._____Is vigilant on a daily basis, observing and correcting and/or reporting safety hazards.

7._____When required, completes employee/property incident investigations with thoroughness and accuracy in a timely manner.

8._____Has reduced recordable incident rate since last evaluation period.

A. Plans for employee development:_____
_____
_____
_____

B. Attitude and commitment toward safety which to a marked degree either adds or detracts from overall performance:_____
_____
_____
_____

C. Significant changes in performance which have been noted since last review:
_____
_____
_____

D. Key areas where performance can be improved: _____
_____
_____
_____

E. Comments on differences of opinion concerning review:_____
_____
_____

This evaluation covers performance for the previous _____months.

Evaluation form completed by: _____Date_____

Evaluation form received by: _____Date_____

I work with companies that have had safety programs in place for ten years where Line Supervisors still complain about the unfairness of it all. Many Line Supervisors still attempt to duck and cover from their responsibilities regarding safety.

Just what should the consequences be for Line Supervisors who do not enforce safety rules and regulations or allow certain employees to get away with violations? What are the consequences for employees who are permitted to violate safety rules? I have used the Supervisors' Evaluation Form (reproduced at the end of this chapter) for years and it has proven to be successful.

Your program will fail without the Line Supervisor's involvement. If safety is a people business, and it is, then every First Line Supervisor is an indispensable part of it.

Let's now take a look at the real people in the process—the line employees.

# 4

# "These People Hate Me."

# The Safety Director and the Line Employee

If you have successfully petitioned to get your boss, the Plant Manager, and the First Line Supervisors on your side, or at least to accept your role, congratulations! There are many in our profession who have not accomplished that during their entire careers!

The whole purpose for your position in the company is the protection of the employees. Look at your OSHA log, look at the injuries occurring to those people—that is why you do what you do.

Let's first determine your history with the employees on the plant floor. Do you know them? Not just what it is they do, but do you know them as individual people, rather than worker bees assigned specific tasks in the hive? Did you come from the plant floor yourself or were you assigned your safety role from the office? Do you have a nodding acquaintance with them as you go through your other daily routine pre-safety duties?

Stop now and think about your relationship to the employees on the line. What is it they know about you? What do you suppose their perception of you is? Please don't say, "I don't have the slightest idea." You have some concept from your dealings with them in the past. Have there been encounters that were negative or positive? What about company activities, such as picnics or Christmas parties? If you were promoted internally within the company, they know you and something about you—you didn't work in a vacuum.

Why the emphasis on the relationship with line employees? Simple—safety is a people business. You weren't just appointed the company Safety Director, but also the company state trooper, town marshal, and chief of police as well. For a while, hopefully not too long, that is the way they are going to see you. It's as if your job is to catch them doing something wrong. That may not be the way you see your role, but that is the way they may see it.

## Two Critical Rules to Follow

There are two rules you need to follow with the line employees at all times regardless of the situation:

1. *Talk straight, play fair.*

2. *Be firm, fair, and consistent.*

Copy these two rules down and put them somewhere you can see them. You will shoot yourself in the foot if you ever break them. The people you are paid to protect will be watching and monitoring your activities all the time and will know whether or not you are sincere about their safety or are going through the motions.

## What To Do First

The reason you were selected as your company's Safety Director is irrelevant now. You're the person responsible for these employees from now on. Get to know them and their jobs. Walk the plant floor from receiving dock to shipping dock and get a feel for the motion of the operations, as well as for the people. Do this for their sake as well as yours. You want them to size you up as a person, as a Safety Director, and as a company rule enforcer. Do this a minimum of twice a day for at least the first three months.

Plant employees are used to routine and habit; get them into the routine of seeing you and talking to you as a Safety Director. You should take a notebook or tape recorder along and jot down notes of things you see or promises you have made. Don't turn this into a formal inspection everytime you walk through. The purpose is to get to know the employees and for them to know you. You should learn employees' names every time you make your tour. But do not force your familiarity on them. If you are from the front office, don't race into the cafeteria at lunch time and try to strike up a conversation with the first table you come to. If you did not eat with them when you were in the office, why now? Don't force yourself on them. Let them come to you as they feel comfortable.

## Where To Start

### Training

The first thing you need to determine is what the line employees know about working safely. Have they been trained in safety? The best place to start is with that training which is mandated by OSHA.

You need to do an audit of just what training the employees have received. There is a reason OSHA is so strict about mandated training. In almost all cases, the lack of training is what results in injuries or fatalities. Training of line employees not only helps them to gain knowledge, it also gives you the credibility you need. You want them to associate you with safety and with safety knowledge. There is no better way to do that than to be perceived as the company's expert on the subject.

Training of line employees is not as difficult as many company owners and safety directors claim it is. Safety training needs two elements to succeed: it must be informative and it must be entertaining. I know that you were not hired to be a standup comic, but if the employees dread attending your training classes, they will have a negative attitude about safety and you.

There are a variety of a fifteen-minute videos that give the safety message you want to get across. Most of them come with instructors' manuals that will assist with the presentation. When the Right-To-Know Standard was first passed, the availability of training programs for it were scarce, and they were boring. Our generation of workers wants to be entertained; use the videos available for that purpose. The key here is to do your homework—be prepared to answer questions should they come up.

Let's stop now and think about what it is we are trying to accomplish. You want the line employees to see you as sincere and knowledgeable in safety. There is no better way to accomplish that than to put yourself in the role of teacher. As children, we grew up knowing that teachers were there to help us, guide us, and educate us. Use that role to start building rapport with the line employees. You may choose to use outside firms for the purpose of training and there is nothing wrong with that. What you need to do is be there to introduce the trainers so the employees see you as the instrument of this education.

There are many excellent books on training available. I strongly suggest you look at some of them for reference. In the next chapter, we will discuss the training that you will need. Later on in the book, we will spend a chapter discussing the "how's" of training the line employees in safety. Please remember, *the single most important thing you can do as the Safety Director is the training of employees.*

## Three Kinds of Safety Directors

We started with the employees' perception of you as a trainer. Let's now work on your recognition as a regulator. You will arrange, control, direct, manage, order, and coordinate their activities regarding safety in the company. The way the line employees will look at you is going to be a reflection of the way you look at them. Do you see them as friends, people you care about and want to protect from injury or death? Or do you perceive them as tools, pairs of hands that perform the scut work for the company? Either way, they are going to know how you see them. Don't kid yourself, they are going to know it, and will reflect back to you the same "front office" contempt that you may be trying to mask. Let's remember again, that *safety is a people business.* You must really care about what happens to these employees whose safety you are guarding.

In the safety engineering field, there are three categories of professionals: *Sailors, Surfers, and Swimmers.* Throughout this book, I will be beating the same theme to death: you must really care about protecting people, have a passion and commitment to it, or find a different job. If you have read this far and have determined that this work is not for you, then you and the employees will have already benefited.

In my many years in this profession, I have run across almost every type of Safety Director there is—from the totally dedicated to the "it's not my job" type. As I said, I can still classify Safety Directors into three categories: Sailors, Surfers, and Swimmers.

Sailors never get wet from the day-to-day exertion of the profession. Oh, they will fill out the OSHA log and attend a safety committee meeting. But they won't get involved personally or professionally in the job. Their skill level is usually very low. They will posture for others in the company, but they could not tell you their incident rate to save their life. They are not committed, caring, or conscientious in the profession. These individuals range from those with degrees and certifications to those who were appointed to the job.

Surfers represent the majority of Safety Directors in the field today. They do get wet and sometimes they even fall into the water. They balance on their safety surfboard hoping to stop injuries, yet are afraid of the consequences of pushing their program too hard. They bob and weave, and sometimes duck and cover, usually doing the best they can with what they have. I have respect for Surfers because they do take risks, albeit occasionally, to try to protect employees and to do their jobs.

Swimmers see their job as a profession and sometimes a calling. Good swimmers are a joy to watch—but they are extremists that can drive a Plant Manager insane by always making demands for the sake of the safety program. That being said, they really care and that alone says volumes about them.

Do you know which classification you fall into?

## Pressing the Flesh—Walking the Plant Floor

To be a successful Safety Director, you must pass among the workers. Get sweaty, dirty, and frustrated just like they do. The first thing you want to purchase is a microcassette tape recorder. Nothing, absolutely nothing, will be more valuable to you than that device. Keep it with you at all times. People are going to be approaching you continually to ask questions, request help, and offer assistance. They are going to gauge you on how well you respond to their needs. The kiss of death for a Safety Director is "I'll get back to you" and not doing it. I told my employees that my tape recorder was my memory. I could remember what they said because it was there on tape. If you establish a reputation as a person who does, in fact, get back to the employees you will be off to a very good start in your role. Remember, you need to go back to them, eyeball to eyeball, to answer their questions. Do not send them a memo; their world does not revolve around memos. Face to face interaction is important for your reputation among line employees.

Employees will ask you questions regarding OSHA's position on something. What I always did was copy the standard and then highlight the relevant parts for them. I would take it to them and offer to explain it if they wanted me to. At first I was being run back and forth, but soon their interest died out. They knew I would tell them the truth. People are going to come up to you constantly asking questions, making suggestions, offering comments, throwing out observations—*Listen To Them!* It is amazing how much they know. Let me give you one example of what will work if you listen to them.

When I first took over a company as the Risk Management Director, I did what I am suggesting to you—I walked the floor with a tape recorder, pressing the flesh, and being seen. One employee rather loudly announced to me and all the other employees that the overhead crane that moved the large rolls of steel had never been inspected and was dangerous. I immediately went to the maintenance department and passed along the concerns of the employee. I was told that "Big Carl" complained about everything, that because I was new on the block he was just testing me. They showed me their inspection and maintenance records, which looked fine to me. Stopping only long enough to congratulate myself on doing my homework, I raced back to the plant floor and gathered all the employees around, including Big Carl, to announce that I had checked on the crane and everything was just fine. I proclaimed in a loud voice that the maintenance people had indeed inspected the overhead crane and that we had the records to prove it. Their response was not laughter, applause, or ever snickers; it was contempt. As Big Carl told me privately, and I will always be in debt to him for that, "Before they just thought you were a fool, now they know you're a fool." He went on to explain to me that the maintenance guy, who inspects the overhead crane once per year, doesn't know the first thing about overhead cranes and admitted as much over a few beers one night. The next day, I called a professional engineering firm to do a full inspection on the crane. When they took the housing off the gear box, broken gears fell ten feet to the floor. Big Carl was absolutely right! Little wonder one American worker gets killed every 99 minutes.

Let's look at this incident. If the crane failed and dropped a load of steel on an employee, killing him, there would be an investigation by OSHA; some, but not much, TV time; some print exposure; plenty of accusations aimed at the company; and probably my firing. There would be immediate comments about "how much the employee would be missed," and how "this can never be allowed to happen again." But, in two months, the employee who was killed wouldn't be mentioned. In most cases, the event wouldn't even make it more likely that a crane was inspected at another facility.

These dangerous scenarios are created by one of two things, or a combination of both: unsafe acts and/or unsafe conditions. What is important to understand is that the company and I would have been responsible and accountable for this employee's life. But other than an OSHA fine there would have been no other consequence for that very preventable accident—none! The widow would at first believe that she could sue until she learns about exclusive remedy, and if OSHA did not cite the company for a "willful violation," she wouldn't have any legal recourse against the company. Her only recourse would be to sue the crane manufacturer and hope to prove negligence. That would be difficult, however, since the negligence occurred with the company, which has exclusive remedy protection from workers' compensation.

In a traffic accident, you can sue for wrongful death when someone is killed, but not in our business. We get to cover up our mistakes with workers' compensation. Workers' compensation is the only recourse open to the victim's family. If OSHA finds cause to cite for a willful violation, companies can often be sued. That, however, is rare. What is missing from this scenario is the accountability of the company, but in the occupational injury/illness arena we don't have that much. I am not an advocate for workers' compensation exclusive remedy reform—far from it. In that case, many businesses in America could be sued every day. What I am in favor of instead is for Safety Directors to do their jobs. Since companies have exclusive remedy, they should assume a more assertive protective role than they are taking. The company Safety Director should really protect his or her employees to whatever extent is required.

What we are about in the safety business is seeing to it that all bases are covered. Regardless of where the information comes from, we should follow the leads to ensure the protection of life. Listen to the line employees; they have first-hand experience with safety every day.

## "What Is That Thing-A-Ma-Jig?" Learning the Operation

Much of what the employees will think of you will be based on what you know about their operation. You will be involving yourself in operations, processes, and equipment that you have never seen or heard of before. It's okay that you do not know; but it's not okay if you act as if you do. The best and fastest way to learn is—*Ask The Employees.* Ask the employee the name of the machine they are working on—not only the nickname, but the technical name. Ask specifically what the machine, operation, or process does. Ask how the part they make fits into the final product? Make notes, listen, and learn. Do this with every operation you don't understand.

Now here is the best thing you can do for yourself and your employees—ask them how they can get injured or killed using this machine, operation, or process. Regardless of your safety background, there are going to be operations and machines you have never seen before. Rather than trying to do research alone in your office on the subject, ask the operators. It's amazing what we in the safety profession learn when we ask for information from the employees themselves. If you approach them with sincere interest and honesty, invariably you will get the same favorable response from them.

## "If You Really Cared About Safety..." Non-Safety-Related Issues

Get ready, Safety Director, the onslaught of anything and everything is about to descend upon you. You'll find that anything and everything under the sun can legitimately be placed in your arena. Can't find a place to park? See the Safety Director. Sandwich machine runs out of food? See the Safety Director. Or, my personal favorite, "Why are we required to drive to work in this rain, sleet, snow, or hail? If you really cared about our safety, we wouldn't have to do this." Hours too long, too short, too much stress, not enough to do, etc., etc. When I train and mentor Safety Directors, I always tell them to watch out for the "nigglies." That's my word for those irritating, annoying, and nuisance type of problems that the employees and other department managers dump on you.

Many new Safety Directors get into the trap of trying to be everything to everybody. There is nothing wrong with that, except that you will probably burn out within the first three months of accepting your position. You can't be everything to everybody, so don't bother trying. All it will do is wear you down, burn you out, and make you negative, bitter, and resentful. Don't do it!

## Is This Safety Related?

During safety committee meetings, this nuisance often comes up: committee members believe that the safety program meeting is the vehicle to resolve all issues, safety related or not. Unless you bring them back to the focus of safety, it can turn into a nightmare of confusion. The best question to ask is always, "Is this safety related?" The response you will probably get is, "It could be." For that matter, any subject can be safety related. A good follow-up question to use is, "Is it covered by OSHA?"

As time goes by, you can respond by simply stating that, "In my judgment, this is not safety related." If you are new to your position, now is not the time to make that statement. It's really a judgment call on your part to determine how far you want to go to restrict employees from fishing for a reason for their complaint to be safety related. My best advice is if you can find it in the CFR, then it will be addressed. Otherwise, it's an issue that needs to be addressed by Human Resources. Do not feel that you are letting the employees down. You need to stay focused on what your job is, not become sidetracked.

# Abe Lincoln Was Right! Trying to Please Everyone Is Impossible

You can't please all your employees—so don't even try. Safety Directors are often seen by employees as approachable—like priests, rabbis, ministers, and friendly street corner police officers. They feel that they can come up to you and speak on any subject because, after all, they see it as safety related. When they approach you with ideas, comments, suggestions, or observations, they are looking for support for their positions. Let me give you an example of how I got into some serious trouble with, "Let's ask the safety guy."

As I made my rounds through the plant, I would stop and chat with as many employees as possible. One day, an employee asked me if I had read the paper or heard the news about a rather grisly crime involving a young child. The alleged perpetrator had been arrested and was awaiting trial. To the best of my recollection, the conversation went something like this: If a little kid can't go to school without being attacked by these monsters, well, then the country is going down the tubes. They ought to just hang that bum, don't you agree?

I was trained in safety engineering, not in politics or diplomacy. I am not a talk show host who smiles and asks insipid questions. I do care very deeply about the tragedy of child abduction and all the rest of the hazards to our children today. My response was: "You find the tree, and I'll get the rope!" I thought that was a pretty snappy comeback. I continued my rounds and forgot the whole incident.

Two days later, my boss called me into his office and asked me if I was aware the union was considering a racial grievance against me. I looked at the man like he was from another planet. "Racial grievance?" I asked, "What are you talking about?" "The comment you made about hanging that child abuser is being construed as a racial remark because you do not want him to have a trial and he is a member of a minority race." "Look, Boss, the child was a member of the same racial minority, and the guy I was talking to is a member of that same racial minority."

The situation was resolved without a grievance, but it taught me something I try to teach to new Safety Directors: be cautious of your comments, because they will come back to haunt you. Employees, especially plant employees, see you as an "expert" on nearly everything and will repeat your comments over and over no matter how innocently made.

There will be times that your answers will be as open and as candid as you can make them with no negative intent. There will be people, however, who will attempt to use you as the sounding board for their personal grievances. Be cautious! The best way to respond is always to picture the company president standing next to you.

## The Bell-Shaped Curve: Learning Your Audience

What is the target population for your safety program? By that I mean out of 100 percent of your employees, how many are you realistically trying to reach? Experience has taught me, and other Safety Directors, about the bell-shaped curve. It is broken down into only three elements, the *plus 10 percent*, the *minus 10 percent* and the *plus/minus 80 percent.*

It goes like this: the *plus 10 percent* are your best employees. Their personal lives are in order; they pay their bills; they have normal, happy, mainstream goals; and they really appreciate the job they have with the company. They feel they have been treated well by the company and the management team. If the company wants a safety program and a Safety Director, it's okay with them. Any company would love to have these employees.

The *minus 10 percent* are the folks, for whatever reason, life has not been kind to, or they have not been kind to it. They complain, moan, groan, snivel, wheedle, and whine. In their opinion, the company doesn't care about the employees, pay is poor, vacation schedule is not fair, parking is inadequate, supervisors play favorites, etc., etc., etc. You would love to give these folks to somebody else, anybody else. They, however, do just enough to keep from being fired.

The *plus 10 percent* you don't need to reach. These folks were on-board with your safety program before there was a safety program. You don't have to reach these people, because they are already there. Your *minus 10 percent,* on the other hand, simply are not likely to be reached by you or your safety program, either. You aren't realistically going to change them, so don't exhaust yourself trying!

Your focus is on reaching the *80 percent.* They are the people who will make or break your program. These folks are the ones who will give you an even break in your attempts to provide a safety program for them. They are not going to love all the rules, regulations, and mandates that you prescribe, but if presented well, they will at least grudgingly accept them. This is the target audience that you want to reach. Remember, *safety is a people business*, talk straight and play fair with them.

In dealing with people in our profession, we must communicate. The next chapter covers what you need to know so you can make much needed technical knowledge known.

# 5

# "I Have To Learn All That?"

# The Safety Director and Technical Knowledge

Let me start this important chapter with an incident from our "Surviving an OSHA Inspection" seminar. One of the participants was very verbal during our presentation. We had an OSHA Compliance Officer and an OSHA attorney also presenting with us. This participant was challenging almost everything everyone else was saying. Halfway through the afternoon session, I began discussing what you need to learn about the OSHA standards and why. My usual opening line is, "You cannot play the game if you do not know the rules." This participant then stated, "I have been a Safety Director for the past nine years and have never read the OSHA standards. I don't even know how! Safety is common sense!" His comments reminded me of the old joke, "Yesterday I didn't know what an engineer was, and today I are one."

What we do not know in our profession will get workers killed! Many new Safety Directors are appointed to their roles without the slightest idea of *what it is they are supposed to do*. Once they have some understanding of what they should be doing, they do not have a clue as to *what it is they should know*. Let's start with that.

## What You Should Know

Which comes first, the chicken or the egg? Or, should I learn the OSHA standards first or the *Accident Prevention Manual*? In my experience in training and mentoring new Safety Directors, I have used both methods. I have reached the conclusion that you should learn how to research the *Code of Federal Regulations* (CFR) first. Don't panic if you don't know what a CFR is. You can't believe how many Safety Directors have no earthly idea what a CFR is, either. The CFR is your guidebook to what it is that OSHA or your state OSHA expects of a safe workplace. The most important thing you need to learn is how to read and research the CFR. Please note: You cannot be a Safety Director responsible for protecting employees' lives if you cannot read, research, and interpret the OSHA standard. Good intentions don't count—you must be able to read, research, and translate. Let's see how it is done.

## Learning the CFRs

One of our most popular seminars is "Researching and Understanding OSHA Standards." It is not just new Safety Directors who attend, we have plenty of "veterans" who claim to be there to "brush up" on their skills. It's not as difficult as it looks.

The first question is, "Where do I get a CFR?" Call your state or federal OSHA, or a commercial publisher like Government Institutes. Simply say that you want to get a copy of the *Code of Federal Regulations*. Be sure that you know which CFR you need. For example, 40 CFR 1910 is for General Industry, and 40 CFR 1926 is for Construction. There are other standards for other types of operations, but for now let's stay with 1910 and 1926. If you are not strictly a construction business, you are probably in General Industry.

Federal CFRs and state plan standards are not necessarily the same books. For instance, California's safety standards come in multiple three-ring binders because they demand so much more than even federal OSHA. Most state regulations are the same size as the federal ones. They are usually 9 inches high, 6 inches wide, and 1 and 1/2 inches thick, but very condensed. The CFRs and state codes contain the standards that cover your requirements and responsibilities in protecting your employees. Do not refer to them as "rules", "regulations," or "codes." Although many of the standards do, indeed, come from rules and existing codes, OSHA refers to them as *standards.*

When you first get your CFR, don't look at it once and then consider quitting—it's not going to be that bad! Open the book to the Table of Contents. Notice how there are Subparts from A through Z? Take a moment and read through them. For instance, Subpart D covers Walking and Working Surfaces, Subpart N covers Material Handling and Storage. Also notice that each subpart and topic has its own 1910 number for your future reference.

The first thing I want you to do is to get two highlighters of different colors. Highlight all the subparts in one color and all the topics in another. If the term "subpart" is confusing to you, then use the term "chapter." That's all that the subparts really are anyway. You will have chapters from A to Z, covering all the different topics contained in the CFR. As you highlight, notice how the topics of the subpart (chapters) are related to that subpart.

Using your Table of Contents, however, is not the only way to reference material on how to research a subject. In all federal standards, and in most state OSHA standards, you can turn to the back of the book and find the index. As an example, look at the back of the book and, using its alphabetical listing, find the heading for "Exits." Notice that under the term Exits, there are topics such as Access, Arrangement, Capacity, etc. To the right of the topics are numbers such as .37(f), .37(g) etc. These are telling you that 1910.37 covers "Exits" in the CFR. To be sure, go back to your table of contents and look up "Exits." Don't see it? Look under Subpart (Chapter) E for Means of Egress. You will find that OSHA is going to be using terms that you may be unfamiliar with.

---

**e•gress**  *n.*  a place or means of going out.

---

In your index you found "Exits" and various numbers and letters such as 37(f). Under the Subpart on "Means of Egress", you will find 1910.37 that covers "Means of Egress—General." That topic covers "Exits." We will return to questions on "Exits," but first let's look at how the number and letter system works.

Let's pretend that your sometimes domineering boss, just came up to you and said, "One of our ungrateful and chubby employees just complained to me that our fire exits aren't wide enough. You're in charge of safety, so find out what OSHA or whoever is in charge of that stuff has to say about it and be in my office in twenty minutes."

Let's show our boss that we can do our stuff. First go to your index. Yes, index. The Table of Contents gives you a good overview of the topics contained, but for a specific item, go to the index. The index states that Exits are in 1910.37. It also mentions all kinds of neat stuff, like access and capacity, but it doesn't say anything about width. Don't worry—you need to have the combination to open the safe. Some indexes give you the page number. If yours does not, then go back to the Table of Contents and look under Subpart E and find 1910.37, Means of Egress—General. Turn to that page. It should say 1910.37, Means of Egress—General. Hold on a minute. Do I hear some whining and sniveling out there? You're doing great! Give yourself credit for using the CFR. As I mentioned before, I know Safety Directors who have no idea what a CFR contains, and then complain because OSHA cites them big bucks when they are not in compliance.

Here we go. You'll need a six-inch ruler or straight edge of some kind, a magnifying glass or a magnifying ruler, and several different colored highlighters. Put your ruler under the heading 1910.37, Means of Egress— General. Now highlight that. From now on, everytime you look in your CFR you will highlight the topic heading. Now slide your ruler down to (a) Permissible exit components. Notice how that is typed in **bold**? Highlight that. Everytime you go into your CFR, you will highlight the alphabetical listings in bold. Keep sliding your ruler down until you get to (b) Protective enclosures. Again, highlight that heading. Now keep going and only highlight the alphabetical headings. Ignore the numbers and roman numerals. Highlight every heading until you get to (f) Access to exits. Highlight that and start running your ruler down to every number until you reach 6. It states, "The minimum width of any way of exit shall be in no case less than 28 inches." Now highlight your answer with a different color so you can immediately find it if you need it, and so you can start keeping track of how much research you have done. You can now go back to your boss and say, "CFR 1910.37(f)(6) states the minimum width of any exit is 28 inches."

The system works like this: Standard (1910.37), Letter (f), Number (6), and Roman Numeral. Also, notice at the bottom of the page the respective CFR number will appear—1910.37. We have not done a Roman numeral yet, so let's try one.

Mr. Yukmeyer is so impressed with your work that he says to you, "Okay, you seem to know what you are doing, so when do I train my employees on the Emergency Action Plan?" Now, open your CFR and turn to the index. Look alphabetically for Emergency Action Plan. Notice the number to the side that shows 1910.38(a). Since you have all those highlighters why not go ahead and highlight what you look up? Now, instead of going to the Table of Contents simply thumb through the CFR, looking at the bottoms of pages until you find 1910.38. If you have trouble, go to the Table of Contents and under Subpart E look for 1910.38.

Get out your trusty ruler and start. Highlight (a) Employee plans and fire prevention plans. Now go through 2, 3, 4, and—wait a minute! What does 5 say? 5 is Training! It covers what your company must do to make sure you have a sufficient number of employees to assist an emergency evacuation. But that doesn't tell you *when* you have to train them. However, the roman numeral (ii) does. It states "the employer shall review the plan with each employee covered by the plan at the following times." Then under your roman numeral (ii) there are the letters (A), (B), and (C) stating when those times are. So again, you can tell your boss that 1910.38 (a)(5)(ii) states when you should train your employees on the Emergency Action Plan.

Let's do one more, then I'll give you some questions (the answers are at the back of the book). How often should your company's fire extinguishers be inspected? Look up fire extinguishers in the index. It shows that the standard is 1910.157. Now flip through the pages looking for 1910.157 until you find the heading 1910.157 Portable Fire Extinguishers. Be sure to highlight your alphabetical headings. Using your ruler and highlighter, keep going until you find inspections. However, you may have noticed in the index that it did say the inspection was under paragraph (e). Try that. Paragraph (e) covers Inspection, Maintenance, and Testing. It says your company will visually inspect fire extinguishers monthly. So 1910.157(e)(2) is your answer. The way you will pronounce this is: "1910 point 157, para e, sub para 2 states the following...."

I will give you some questions to look up to give you some practice in this area:

1. Which standard requires that cylinders be examined to determine a safe condition?

2. What are the training requirements for employees regarding fire extinguishers?

3. What is the safe distance for following another forklift specified in the standard? (Hint: OSHA calls forklifts "Powered Industrial Trucks.")

4. How is "Point-of-operation" defined by the CFR?

5. What is the standard requirement for work rests on abrasive wheel machinery?

6. When an employee requests access to medical records, what are the requirements of the employer?

7. How is a stepladder defined?

8. What are the general requirements for eye and face protection?

9. What are the written requirements for respirators?

10. What standard requires certification records of each inspection conducted for a powered platform used for building maintenance?

## Elephant Stampedes Are Under Which Standard? The General Duty Clause

I have been told by a veteran OSHA Compliance Officer that the CFR 1910 General Industry Standard has over sixty-five thousand standards in it. You would think it covers just about anything you would want to know, and probably more than you would ever care to know. Not really. Right after the passage of the legislation enacting OSHA, new compliance officers hit the street to make a "Work Place Free of Recognized Hazards." One officer ran into an interesting situation. It seems that he was inspecting a construction site when he observed an employee on the second floor of a building throwing large pieces of concrete out of a window into the back of a pickup truck. That wasn't so bad, but there was another employee in the back of the truck dodging the concrete pieces as they came hurtling down. The Compliance Officer cited the company for this procedure. The employer, however, took the citation to a hearing and challenged OSHA on which standard they were citing him under. Unfortunately, OSHA did not have a standard entitled, "Really dumb and dangerous things you should not do." The hearing judge had to throw out the citation because there was no standard covering the cited practice. OSHA then decided they needed a standard or clause that covered anything not already in the CFR. Thus was born the General Duty Clause. The General Duty Clause was written to cover those situations where there are no specific written standards.

Under the General Duty Clause:

1. The employer shall furnish to each of his employees employment and a place of employment that are free from recognized hazards that are causing or are likely to cause death or serious physical harm to his employees.

2. The employer shall comply with occupational safety and health standards promulgated under this act.

The key to the General Duty Clause is the term "Recognized Hazards." A recognized hazard is a condition that is "of common knowledge or general recognition in the particular industry in which it occurs and is detectable by means of senses, or of such wide, general recognition as a hazard in the industry that there are generally known and accepted tests for its existence that should make its presence known to the employer."

I have been asked several hundreds of times for my interpretation of the General Duty Clause. I believe that the phrase "detectable by means of senses" says it all. The most important rule to follow in your role as Safety Director is, *When In Doubt, Check It Out!*

There were three types of business owners: catch me if you can; just enough to get by; and safety smart. Well, the business owner of the pickup truck incident was defiantly "Catch me if you can." He wasn't just in violation of a safety standard, he was also guilty of being "Felony Stupid!" It really is the responsibility of the employer to make a good faith intention to protect the employee—and it's most certainly your job as the Safety Director.

Okay, we now know what is expected of us, so what do we study to be able to acquire the knowledge to do it?

## Safety Training—Discernment Is the Key

Every day, companies send me brochures, flyers, letters, and every other kind of correspondence telling me how great their safety training, information, and assistance programs are. Many of them want my business and/or endorsement. Very few get it. I am going to pass on the best that I have encountered or discovered in the past thirty years.

## National Safety Council

Nothing beats the National Safety Council for putting out the very best beginner's safety training. I have no business or financial relationship with the Council, other than recommending their publications to every new Safety Director I train and mentor. Locate your local chapter of the National Safety Council in the phone directory—nearly every state has at least one chapter located in the state capital. If you cannot find a local chapter or state chapter, call NSC's toll free number at their headquarters in Itasca, IL (800-621-7619). Ask them to send you a catalog of products. The catalog is free and it will show you what is available.

# Books to Buy

The first book you need to purchase is the National Safety Council's two-volume set *Accident Prevention Manual for Business & Industry*. Also buy the *Study Guide* for the *Accident Prevention Manual*. The manual is in its tenth edition, and anyone who is serious about the prevention of occupational death and injury has read it at least once.

Volume One covers Administration and Programs, with twenty-one chapters covering such topics as hazard control programs; occupational health programs; acquiring hazard information; accident investigation, analysis, and costs; and attitudes, behavior, and motivation.

Volume Two has twenty-three chapters and gets deeper into the engineering aspect of your profession, with chapters covering such topics as buildings and plant layout, ergonomics in the workplace, fire protection, personal protective equipment, and flammable and combustible liquids.

I would suggest reading only one chapter per week and then answering the corresponding questions in the *Study Guide*. The answers are at the back of the guide. If you do only this, you will probably be better informed than at least half the practicing Safety Directors on the job today!

Industrial Hygiene is another subject you need to know. Industrial hygiene is defined as, "The science devoted to the recognition, evaluation, and control of those environmental factors or stresses that may cause sickness, impaired health, or significant discomfort to employees or residents of the community."

The National Safety Council's *Fundamentals of Industrial Hygiene* is one of the absolute best resources you will find. Its seven parts and thirty chapters cover areas from industrial toxicology, industrial noise, ergonomics, and local exhaust ventilation, to name a few.

Don't study industrial hygiene until you have completed reading your accident prevention manuals. Then you will understand why you need to have a least a working knowledge of industrial hygiene.

Another very good book to buy and read is the *Complete Manual of Industrial Safety* by S.Z. Mansdorf, published by Prentice Hall. This excellent book has twenty-six chapters covering everything you will need in your role as Safety Director.

Government Institutes is another very good source for safety and health publications. Their books are written by professionals in the field—those who are actually doing the work. Get a catalog from Government Institutes (call 301-921-2355) and see what they are offering—I think you will find them very helpful.

I recommend starting with these books. They will give you the best all-round knowledge that you need to perform your job. Let's talk now about formal safety training.

## Register Now, Seating Is Limited!

What about formal safety training for you? There are certainly enough people, firms, and organizations claiming they can make you a safety engineer overnight—but they cannot. I can't, either.

The safety profession is no different from any other. It has its questionable practitioners, too. I can't begin to tell you about the retired, fired, or laid off police officers, plant managers, military personnel, salesmen, clerks, and former safety committee members who try to pass themselves off as safety professionals. Just because I was once a patient, doesn't make me a doctor. Be very, very careful of these people. Most of them haven't a clue as to what they are doing. Do not trust them—ever!

Let me tell you a quick story. A retired supervisor from a steel mill on the East Coast attended a "So You're the Safety Director!" seminar. He was currently working as a salesman for a safety training firm. He came to our seminar so he, in his words, "could learn something about the safety profession." He took our one day seminar and then began calling himself a Safety Consultant. The term for what he was doing was fraud, pure and simple. Always be cautious of those who claim to be safety professionals.

The best place to contact for safety training is your local chapter of the National Safety Council. They provide the safety training you need with instructors who have proven credentials. The National Safety Council has a vested interest in seeing to it that you receive what you need—it's called ethics. I do occupational safety training in association with our NSC local chapter. We use their approved training programs, materials, and outlines. The NSC should always be your first call. You can also call your local Employer's Council and ask if they are offering any courses or whether they will offer a referral. Please remember not to accept just any program or referral.

## "We're OSHA Approved"

I get at least three calls a month asking me if I can make someone OSHA approved. I am not OSHA approved and there is a good reason for it—*OSHA does not approve anybody or anything!* I was once conducting a seminar when a participant argued vehemently with me that OSHA does indeed approve safety engineers and products. I immediately called federal OSHA in Washington, DC, and they faxed me a statement saying that they do not approve anything—no people, no products! Those vendors who sell safety products love to put "OSHA Approved" on their products because it gives the buyer confidence in what they are buying. But, OSHA approval does not exist. This includes the training you are considering. If the organization or individual claims they are "OSHA Approved" ask them to get it in writing from federal OSHA in Washington. The next sound you will hear is a dial tone.

## You Too Can Be Safety Certified

The next point of confusion for the new Safety Director is certification. Again, questions and phone calls from individuals come to me with requests to grant certification. Since I am not a certifying organization, I cannot do that. It takes years of training, education, and experience to meet the stringent standards of a recognized certification board. Taking a ride in an airplane will not make you a pilot. The safety profession is just that—a formal profession with a structured discipline.

When Safety Directors ask to be "certified" after training, I offer them a sample examination from one of the certification organizations. That usually stops all further demands for immediate certification.

The first credential I suggest for new Safety Directors is the National Safety Council's Advanced Safety Certificate. To earn this certification, you will be required to take a prescribed course of seminars, including Accident Prevention, Industrial Hygiene, Safety Training, and Safety Management. It is well worth the time and expenditure to earn this recognized certification.

The two most recognized safety organizations offering certifications are the American Society of Safety Engineers (ASSE) and the World Safety Organization (WSO). The ASSE is much larger and older than the WSO. The ASSE offers certification as an Associate Safety Professional (ASP), which is their first step towards the Certified Safety Professional (CSP) designation, their highest.

There is no question that ASSE is better known outside of the safety field than the WSO. The WSO is a new organization, about twenty years old, and is very inclusive for all members regardless of knowledge or experience, encompassing a world membership. The WSO is growing very rapidly, especially with in-house appointed Safety Directors. The WSO is also chartered with consultative status to the United Nations.

If you are interested in certification and are new in the profession, I suggest the World Safety Organization's Certified Safety Technician certificate. You will not be required to take a formal board examination and it is geared for those at the entry level. After that, you will be required to take difficult examinations, requiring many hours studying, from either organization to receive additional certification.

Both organizations' written exams are thorough and equally difficult, contrary to what you may hear. The whole idea is to learn about the profession. But be aware that only 12 percent of practicing in-house Safety Directors are certified by an organization. The other 88 percent are doing their job every day and either do not know or do not care about the certification frenzy.

I recently had a young man, with three years in the profession, ask me how he could get his certification so he could become a consultant like me. He was wondering if he should take advantage of those seminars that are offered in safety publications to help him pass his CSP. There are firms which, for a rather large fee, will tutor you for taking your boards. I have real difficulty respecting these "Stuff and Spew" seminars. It reminds me of microwaving a turkey. Safety Directors need to be cooked with experience and basted with knowledge. There simply is no other way to learn the job.

I furthermore encourage you to get your certification for the *right reason.* If you want to be part of an organization that you can network with, or that will offer seminars and conferences, then, of course, join a safety organization. If, however, you feel that your certification will give you more knowledge or expertise or even, as some members feel, a higher status in the field, you will be sorely disappointed.

If you work in a small company that does not require certification in the safety field and you have no burning desire to obtain one, then I suggest you do not do it. It is very time consuming and expensive.

If you do want certification, then contact both the WSO and the ASSE for their certification process. Please, understand that a certification will not make you a better Safety Director. It will simply make you a certified Safety Director. *Certification does not mean qualification!*

## To Certify or Not? A Case Study

Allow me to tell you about a remarkable individual in our profession. In the dedication to this book, Kevin Wilt is mentioned. Kevin is one of the 88 percent of Safety Directors who simply do their job and are not obsessed with certification. I first met Kevin in November of 1991, when he had been selected as the company Safety Director from his position as Warehouse Supervisor. He had no knowledge or experience in the safety profession and was rather doubtful that he had made the right decision in accepting the promotion. Kevin was eager to receive my mentoring and training.

In November of 1991, Kevin's company's incident rate was 115. There was no program to speak of, so he and I created what was needed. With no college background or technical experience, he reduced the incident rate to 10.7 in four years. Though his company was almost out of options for workers' compensation insurance in 1991, they are now self-insured. OSHA had previously made a wall-to-wall inspection with serious violations, but now the agency comes to his plant to study, not to inspect, his operation. Though that plant was once considered dangerous, recently they have won the Governor's award for the safest large business in their state for three consecutive years. This is a company with over one thousand employees with shifts working seven days per week, twenty-four hours per day.

Kevin Wilt is an excellent example of the unheralded 88 percent who do their job and take pride in it. He was the best student I ever had. He is without a doubt the most knowledgeable Safety Director I have ever met! He instinctively and intuitively knew how to apply his knowledge for realistic improvement. There wasn't a book or article that I gave him that he didn't devour and attempt to implement the concepts of in his company.

During the five years I had the pleasure of training him, I often suggested certification. His response was always, "Will it protect any more employees, or reduce my incident rate?" The best way for me to describe this true safety professional is to use his own words. "You can't do the job for the paycheck, it must be from the heart." Kevin has now been promoted to his company's Chief of Plant Engineering. Kevin hand picked his successor.

I am now in the process of training this successor and am happy to report his motivation and dedication is equal to that of Kevin Wilt. Safety truly is about people, not certification. Granted, not all of the "silent 88 percent" are as dedicated as Kevin Wilt. But in my experience, there are infinitely more dedicated people in our profession than we realize. Day after day, they slog through the swamps of angry plant managers, resentful supervisors, and disgruntled employees and keep their focus on creating a safe work environment. My admiration for these true safety professionals knows no bounds!

## Realistic Certification

I want to be very clear on the subject of certification: I am in favor of it for the right reasons, especially in the consulting field. Our profession needs a clearinghouse to sort out unqualified practitioners. I am confident that there will eventually be realistic national certification standards that are approved and accepted by all safety organizations. Even then, the certification will not make anyone more qualified, it simply will eliminate those who do not have the motivation, commitment, and initiative to learn our profession. What I recommend for certification is technical knowledge and years of experience. The technical knowledge required for the safety profession may be obtained from many national organizations, safety councils, or universities and colleges, but no certification can replace real on-the-job experience. One safety professional has over thirty years experience and more technical knowledge than anyone I have met. For him not to be recognized as certified because he hasn't taken an examination, versus a Loss Control Inspector with five years experience who has passed a test, is absurd! Once again, *certification does not mean qualification!*

## Hub Cap Safety: Feel Good Programs

In your role as a Safety Director, mail will be coming across your desk at an alarming rate, all of it informing you on how you can stop employees injuries and save huge amounts in worker compensation premiums by adopting their program. If that were only true! There are excellent programs available to you that you should consider for the right reason. Unfortunately, many new Safety Directors use programs for the wrong reason. The National Safety Council, DuPont, CLMI, to name a few, offer superb programs that will help you in your efforts to improve company safety.

I referred earlier to programs that don't really help you in the "here and now"—called hub cap safety programs. As mentioned earlier, hub caps make your car look better, but they certainly don't make it run any better. Likewise, behavioral safety seems to be the safety concept in vogue now. I'm glad to see that, because I'm as much a behaviorist as an engineer. But, be sure you know what you need, not just what you want. Do not pick a program because everyone else is using it and it comes highly recommended.

Conduct an inventory of what it is you need to reduce injuries. Where are you injuries occurring? Why, in your estimation, are they occurring? What needs to be fixed? If you are unsure of either, call an outside source, including an OSHA Consultative, for assistance. The important thing to focus on is, "Will this program protect employee and reduce injuries?"

Let's now spend some time on a very important person in the process—you!

# 6

# "Woe Is Me!"

# The Safety Director on the Safety Director

"Woe is me! I try and try, but no one listens to me. How can I fix this place if they won't let me do my job? No sense in seeing my boss, he (she) will only say no, or make me feel stupid. Woe is me."

## Hitting Your Ceiling

You are going to hit your ceiling; all Safety Directors do. It's what you do *after* you hit your ceiling that determines what type of Safety Director you are.

As you start your journey to protect people and property, the task at hand will at first seem interesting, then overwhelming, and finally it may become drudgery. It will also start to dawn on you that many of the promises made because management "cares" about their employees are not being kept. Little by little, you will learn the limitations of what you can really do to improve your safety program. You will see where additional training is needed, where a better type of guard will do the job, or where more advanced seminars for you will enhance the overall efforts.

Please keep in mind that as a conscientious Safety Director, frustration will be your constant companion. There is a correlation between your commitment and your frustration. So what are you going to do about it?

## Just Who Are "They" Who Lack the Support?

In our seminars, many Safety Directors express how they are not appreciated, supported, or respected. What is to be done about that? Should you just keep plodding along angry, frustrated, and bitter? No! The first thing I always recommend is to find out *who* is not supporting you? *Who* is not giving you what you want or need, and *why* aren't they?

Let's start with the who. Stop right now and make a list of those people who are not supporting your efforts in the safety program. Is it your boss, your boss's boss, or the company president? Maybe it's the Plant Manager, the Maintenance Director, or some or all of the First Line Supervisors. Be specific as to what it is they are doing that they should not be doing, or what they are not doing that they should be doing.

Whenever I ask seminar participants to be specific about their complaints, they get vague and ambiguous as to what it is that they are frustrated about. If Safety Directors are, indeed, in the business of protecting lives and property, they should not be spinning their wheels playing "Ain't it awful?" Rather they should find out what is wrong and what is needed to correct it. So, let's look at some ways to help you accomplish what it is you want to accomplish.

The first thing you want to do is to document exactly what it is that you need that you are not getting. Don't whine, snivel, or cry—just conduct yourself professionally. Explain to your boss in detail what it is that you need. Here is a common complaint from our participants: "I can't get supervisors to enforce our safety rules and regulations. When I find something wrong in their department they nod, but nothing gets done."

Okay, try this technique.

# MEMORANDUM

**DATE:**

**TO:**   Your Boss

**FROM:**   You

**SUBJECT:**   Lack of Safety Compliance in Department #23; Jim Williams, Supervisor.

1. For the past two months, Department #23 has had a total of eleven eye-related injuries. That is 46 percent more than any other department in the company.

2. I have had three informal meetings and one formal meeting with Jim Williams, the department supervisor, regarding this issue. (Please see attached memo.)

3. In each of these meetings, I have told Jim Williams that I have observed his employees working without eye protection, which is in violation of our company policies and OSHA standards. He has promised that he will enforce our eye protection policy to the letter. On subsequent visits to his department, however, I again have found employees without eye protection. On each instance, Jim Williams has been courteous, attentive, and supportive of his responsibility in the enforcement of the policy to me. Yet he has not demonstrated by his actions that he will enforce the needed compliance.

4. The attached cost breakdown reveals that the eleven eye injuries have cost the company a total of $1,287.34. This cost will be passed on to us through our Workers' Compensation premium.

5. The failure of Jim Williams's enforcement of our policy on eye protection is a direct and citeable violation of OSHA standard CFR 1910.132 and CFR 1910.133. This violation can cost us up to $7,000 per occurrence. Also, if this is cited by OSHA as a willful violation we can be sued by the injured employee regardless of the exclusive remedy of Workers' Compensation.

6. I would like to have a formal meeting with you to discuss and plan a course of action to solve this serious and continuous problem.

cc: Jim William's boss
    Jim Williams

In an earlier chapter, we discussed RACK—Responsibility, Accountability, Consequences, and Knowledge. Does Jim Williams have knowledge of the company's policy on the enforcement of eye protection in his department? If he does, he is not using the knowledge. Is Jim Williams responsible for the safety of his employees, and responsible for assuring their compliance with the company's policies and procedures on eye protection? If so, he is apparently failing in that responsibility. Is Jim Williams accountable to the company, his boss, and to his employees for their safety? Yes, he is and, again, he is apparently failing in that area as well. One consequence for his nonperformance should be having the safety director address this issue immediately. When offered this course of action, I often hear "If I do that, the supervisor will make my life a bad dream. I'll get no support or respect at all." My response always is, "What kind of support are you getting now?" "What kind of respect is he presently giving you now?" What is more important, whether the supervisor will like you or whether you are assuring that your employees' eyes are protected? As the Safety Director, it's not a question of not knowing what to do, but rather having the moral courage to do what you need to do to fulfill the responsibility of your job.

I also hear from clients and seminar participants, "You make it sound so easy. It's not like that in the real world, at least in my company." First of all, I'm aware that it isn't easy at all! It's very difficult for some people to be assertive and even, demanding. It's more difficult to complete an accident investigation on an employee's loss of an eye and explain why this preventable injury happened than to be forceful in preventing an unsafe situation that will allow the injury to occur. Let's be clear about this: being a good, committed, and caring Safety Director is not and never will be easy. Let's now go to another scenario—dealing with your boss.

Here is a real world example of what happened with a client of mine in the Pacific Northwest that illustrates how Safety Directors can get so enmeshed in what's not working that they cannot see solutions. The Safety Coordinator had been wanting to start an ergonomics program in her company for some time. Her boss, the Director of Operations, had allowed some steps in that area, such as job rotation, but had made no real commitment to it.

During a recent safety committee meeting, the issue of ergonomics came up. The Director made the comment that he could put forth a good argument against ergonomics. The Safety Coordinator sat in the meeting and just seethed, making no comment. After the meeting, she said to me, "See what I mean? No support at all. What do you think the other committee members think when there is no support from the Director?"

First of all, let's consider what really happened and why. Please understand this Safety Coordinator is more in tune with the hot buttons of the Director than I am. She just refuses to acknowledge them. The Director is really concerned about just one thing—production! It doesn't make him an evil person—it makes him a focused person. He has given the Safety Coordinator much support.

The Director is well aware that sooner or later there will be a OSHA ergonomics bill and he will be mandated to accept it. Until then, he wants to put it out of his mind. Why? The reason is simple, his technical advisor on such issues, his safety coordinator, has not been doing her job of slowly educating the Director on this issue. For her, it's either 100 percent ergonomics or nothing at all—black or white.

The Director had the same reaction years ago to the Americans with Disabilities Act. His first reaction was, "It's going to put us out of business." Of course, it did not because in that case he was educated slowly by the person responsible for its implementation. During the safety committee meeting, I leaned over and whispered to the Vice President that I would get a copy of information on ergonomics so he could see an example of what he would be dealing with sooner or later. Why couldn't the Safety Coordinator have done the same thing? She was just so frustrated about the subject she could not see another way to approach it. This same Safety Coordinator later told me how frustrated she gets with her job because, in her words, "It's never done."

No, Safety Directors, your job will never be done! There is always something out there waiting to happen. Your job is *never* done; there will always be frustration and more work to do. If this is causing you that much difficulty in your role, then you should consider getting out of the profession. Let's now look at the every popular complaint of, "I just don't have the time, yet I'm expected to get things done!"

## "Do the Best You Can."—Educating Your Boss

I must hear that quote a thousand times a year during our seminars and training of Safety Directors. To a part-time, or even a full-time Safety Director, it's one of the most patronizing and frustrating statements a boss can make. First of all, just what is the "best you can do?" Have you ever really thought about it? What if you had the time to do what you wanted with your program? How would you improve it? Does your boss know this? Have you ever really told him or her? It has been my experience with many bosses that they haven't the slightest idea what it is you do now, not to mention what it is that you could do if you had the time and resources.

Quite frankly, it's not the boss' job to know. It's your job to educate them as to the realities of the role of the Safety Director. Please keep in mind that bosses have a general idea of your role and, in many cases, see safety as a necessary and expensive outlay from their bottom line. It stands to reason, then, that the less you do, the fewer problems you will bring them and the lower expense to the bottom line.

First and most importantly, you must determine what it is you are really supposed to be doing in your role, and how much time it will realistically take you to do it. You will be surprised how much your role demands once you start to do a time management analysis on it. Write down what your safety requirements are.

| | |
|---|---|
| **Training:** | OSHA mandated; non-OSHA mandated, but safety related; supervisors' safety development. |
| **Inspections:** | Daily, weekly, monthly, quarterly. |
| **Accident Investigations** | |
| **Incident Investigations** | |
| **Workers' Compensation:** | Return-to-work; contact with injured employee; insurance company; physician, therapist, miscellaneous. |
| **Report Writing:** | For all the above. |

I have only given you a microscopic amount of what you do or should be doing. Now is the time for you to really break it down into time segments. Please do not say, as so many Safety Directors do, "I don't have the time to determine how much time it takes." The question here is, do you want to be able to show your boss what it is you do and how much time it takes, or do you just want to complain about it? Is it a defense against doing a more thorough job or doing the job at all? Don't get mad, get efficient!. If you are sincere in your role, as are so many part-time and full-time Safety Directors I have met, then it may mean you will have to do some of this on your *own* time if you are to make your point with your boss. Let's use a sample memo on this subject.

---

## MEMORANDUM

**DATE:**

**TO:**　　　　　Your Boss

**FROM:**　　　　You

**SUBJECT:**　　Time required for safety activities

1. Please find the attached time breakdown sheet on safety-related activities requiring my participation. A review of this will indicate to you that I am presently spending _____ hours per week on safety work.

2. Attached also is a time breakdown sheet on my full-time position as Human Resources Director. As you can see, I am spending _____hours per week on this activity.

3. Both my full-time position and part-time position are being neglected and the company is not being served adequately by me in either position due to lack of time.

4. I would like to meet with you to discuss some possible solutions to this problem.

---

Before you send the memo to your boss, visualize him or her reading it and the anticipated response. Think about your relationship with your boss. How does he or she view you? What is the common reaction to you when you bring up safety-related issues?

The initial reaction of many participants is "I have no idea how my boss views me." My response is always, "Nonsense!" You know, at some level, how your boss reacts to you and views you; you can't help but know. The real question is, are you willing to take the time to present your information to him or her in a positive and professional manner?

What is your boss's interaction with you? Is it patronizing, sarcastic, arrogant, indifferent, amused, or just plain rude? Whether he or she is difficult to convince or not, the facts still remain facts, sooner or later he or she will have to admit to them in or out of your presence.

Read the following boss' responses, before you prepare your memo, so you can anticipate your boss' response. Are you tired of being shuttled around and treated with disrespect? Then do something about it. Not only for yourself, but for the people you are charged with protecting.

## The Boss's Dozen Responses

These are probably what will be going through your boss's mind:

1.  Oh, give me a break! Those times are grossly inflated.

2.  Safety doesn't take that much time.

3.  Look out, he (she) wants a raise!

4.  Why can't he (she) pull his (her) weight like everybody else?

5.  If he (she) thinks I'm going to hire another employee, he's (she's) nuts!

6.  Why can't he (she) just stop whining and do his (her) job?

7.  I wonder what's really bothering him (her).

8.  I'll call one of my golfing buddies; his HR Director handles safety and doesn't cry about it.

9.  If he (she) wants the job, he (she) will just have to make do.

10.  What's the worst that can happen if things go along as they are?

11.  Who else can I get to replace this sniveler?

12.  I'll just tell him (her) to "do the best he (she) can."

## Ten Comments You Don't Want to Hear

Obviously, I do not know your boss and what the reaction will be. However, you do need to think about the response that will be made and your rebuttal to it. If you are like some of the people in our seminars, your answer to me might be, "I know what my boss is going to say—do the job or seek employment elsewhere." In other words, you're sure you will be fired for raising this issue. If that is the case, what do you think your boss will say when there is a serious injury or severe OSHA citation and subsequent monetary fine? How about:

1.  Why didn't you tell me?

2.  Were you aware of this? If not, why not?

3.  My door was always open to you.

4.  Didn't I tell you that I supported the safety program?

5.  What have you been doing all this time?

6.  If you needed more time, why didn't you say so?

7.  I thought you were responsible enough for this position.

8.    I am very, very disappointed in you.

9.    Couldn't you have done anything to prevent this?

10.   Under the circumstances, we are going to have to let you go.

Now, is letting your boss know about your lack of time so difficult to do? Do you really have a choice? If you don't raise the issue, you will eventually lose your job anyway. What about the people you are supposed to be protecting?

Writing this chapter was not easy for me because at first it appears that I am callous and heavy-handed towards the frustration of Safety Directors. Not so! I hurt for them when I hear their problems time and time again at seminars. Sooner or later, you are going to have to bite the bullet and do what is best for yourself as well as the employees of the company. If you find yourself in this position, then "talk straight and play fair" with your boss, yourself, and your employees. If your boss will not or cannot listen to your frustrations and problems, then you will have to determine what you need to do.

Let me tell you a quick story about a young man who went through much of what we have been discussing. He came from the east coast to the southwest to take over the role as a company Safety Director. He was promised much and given little. His company has around 700 employees working in metal fabrication. After an initial risk assessment, he determined that the company was not in compliance with many OSHA standards and that their incident rate was almost double the national average. He was told, "That's why we hired you." When he asked for funds for corrections, he was told he would eventually get them.

During his first week at the company, OSHA did a wall-to-wall inspection. In the first forty-five minutes, they cited him twenty-nine thousand dollars for violations they discovered. He was panic stricken and called me. I referred him to a good OSHA labor attorney. They were able to reduce the fines by 25 percent.

The Safety Director was now positive that he would be given the funding to do his job. Over a six-month period, he was given promises, but no money.

Finally, OSHA came back for a re-inspection and cited them again for failure to abate. The Safety Director was fired to show the company's "good faith" in correcting the problem. During this six-month period I often spoke with him on the phone. Over and over again, he would tell me that no matter how he tried to persuade, convince, or reason with management, they just kept promising and not delivering what he needed. When OSHA finally returned, his company fired him and hired another Safety Director. Now he is working as a security guard at a shopping mall while looking for another safety position. He called recently and told me that he should have taken my advice to start looking for work elsewhere. He just kept thinking that management would eventually support his efforts. I admire his commitment and dedication, but regret he didn't see what was happening around him.

## Two Games To Stay Away From

The two biggest games people usually play at work are: "Ain't It Awful" and "Yes, but." It has often been said that the only good job is the one you just left or the one you are starting. Safety Directors suffer the same flaws as any other human being in the workplace.

Safety Directors, however, feel so isolated and misunderstood that they sometimes play these two games to excess. "Ain't it awful?" is simply sitting around telling yourself and others how bad everything is. People will usually be courteous at first and then begin to run from you like you have the plague. Playing "Ain't it awful?" only reduces you and your position in the eyes of others. Don't do it! Every department and manager has his or her own tale of woe to tell. When you play "Ain't it awful?" however, your comments can often be misconstrued to the point that people may believe that they could be killed at any moment. That is usually not the case and it decreases your professionalism.

The other game that has a lot of players is, "Yes, but." This game, or variations of it, involves asking for help and then responding to someone's suggestion by stating, "Yes, but..." "Yes, but, we can't do that because...." "Yes, but, my boss won't let me." "Yes, but, the employees won't do it." "Yes, but, there is no money." "Yes, but, no one respects me." "Yes, but," "Yes, but," "Yes, but." I encounter this game often in seminars when I am asked to resolve a problem. Good players have gone as high as a dozen "Yes, buts," before I allow them to win. It is simply a confirmation that you either cannot or will not find a workable solution on your own. Sooner or later, people will become weary of listening to the game and run from you. Don't play that game!

When you have done your homework regarding support and commitment from management and you realize that it simply isn't there, then it's time to look for another position. That position may or may not be in the present organization. Remember, before you play "Ain't it awful?" "Yes, but," and "Woe is me," try the techniques I have recommended. They have worked for others; hopefully they will succeed for you.

Let's now move on to a very difficult subject—Safety and Employee Discipline.

# 7

# "Let's Not Get Carried Away With This Stuff."
# The Safety Director and the Disciplinary Process

When you first accepted your position as Safety Director, you never had any idea how difficult it was going to be to convince people of the need for safety. Safety is something like good dental health and Mother's Day—everyone is in favor of it. How could anybody not want to work safely? The vast majority of the working people of this country want a safe work environment; it's just that they want it on their terms, not necessarily on yours.

A very big frustration of the Safety Director is not only getting people to work safely, but holding them accountable when they don't. Safety Directors do not enjoy wearing the "black hat" in their roles. Many Safety Directors see themselves as the company chaplain. Safety Directors are supposed to be "safe" people to approach, like the small town police officer or the local clergy.

In many cases, employees see the Safety Director not so much as a part of management, but rather an emissary representing management's interests. In many cases, Safety Directors promote this role to assure their "one-of-the-gang" membership with the line employees. They see themselves above the mundane role of holding employees accountable for not following safety rules. Rather, they see themselves in the lofty world of observing work behavior and notifying supervisors of their employees' misdeeds. That way their hands are clean, their consciences clear, and their popularity secure.

Let me be clear on this subject: Supervisors *are* responsible for enforcing the rules and regulations of their employees. Safety Directors should see to it that Line Supervisors do their job! Sometimes the Safety Director is going to have to be personally involved with the day-to-day people problems of running a safety program. Remember, *safety is a people business.* It's all right to hold supervisors accountable and insist that they hold their employees accountable. It is not all right to pass them off to personnel or other offices when the Line Supervisor is trying to get assistance in how to discipline the employee.

You need to be there for the Line Supervisor and any other manager when it comes time to discipline employees. You are not an objective observer, but rather an active participant in making sure that everyone follows the safety rules.

Another thing to consider is that employees are going to see through your disguise as one of the good guys and you will lose any respect you would hope to have. Again, talking straight and playing fair is the only way to operate in your role.

If playing Mr. Rogers to your employees is ineffectual, then imagine being Rambo with them. There are just as many storm trooper mentalities as there are good guys. It seems that the middle of the road position is a difficult one to take. When you took over as Safety Director, your employees expected you to be a police officer with an attitude. So, it will not be difficult for them to see you in that role when you issue a safety proclamation. They will be watching for facial affect, body language, and tone of voice.

I have seen some of the worst mistakes made by Safety Directors because they felt they were ordained by a higher power to right the safety wrongs of the world. Usually only two things happen to them: They are so disliked and resented they do not last, or, with no support from management, they are considered a joke. Either way, they are not protecting people. I know you're tired of hearing, "talk straight and play fair," but it's really the only way to be successful in this profession that deals with other human beings.

## The "Linkage" Mentality

When addressing a group of Safety Directors on the subject of the disciplinary process, it has always astonished me that they become very uncomfortable with the subject. It is as if discipline is really out of their province and they would much rather discuss machine guarding or threshold limit values. What has brought us to this sad state where tardiness, theft, insubordination, and drug/alcohol abuse will result in discipline, but not safety rule violations? I think part of it comes from what I call "linkage mentality."

Linkage thinking goes something like this: "Yes, I know Bob violated our safety rules on eye protection, but he did have to go to the hospital with an eye injury. Isn't his pain and suffering enough for you? Does he really have to get a disciplinary letter as well?" That logic would lead us to the following, "Even though Mr. Jones did go ninety miles an hour in an occupied school zone, he did run into a tree and sprained his neck. His injury is enough punishment for this crime and no ticket should be issued." Now, before you jump on me about safety violations not being a crime, just stop and look at the blurred logic that is going on in a large portion of American industry.

For some reason, companies—and I do not automatically blame Human Resource Directors for this—seem to be reluctant to separate the injury from the injured. Why is this? What caused the injury? An unsafe

**95**

act or an unsafe condition? If the employee committed an unsafe act, resulting in an injury to another employee, what would our reaction be? What then, is the difference if that same unsafe act resulted in an injury to the perpetrator of the unsafe act? The sooner Safety Directors accept that employees are responsible for their actions regarding safety rules, the sooner we can start enforcing our mandates on safety conduct.

How, then, do you start looking at safety rule violations differently?

The first thing you want to do is take a long, hard, serious look at how many, if any, disciplinary violations have been given by you, Human Resources, or anyone else in the company for failing to follow safety rules. Remember, you should be holding Line Supervisors accountable for their employees and it's important that you help clear the road for their efforts. How should you and/or Line Supervisors document safety rule violations? What is your present policy and is it really being enforced?

Be cautious in this search for the truth because you may run into the neutral response of, "We always discipline our employees for safety rule violations, just like we discipline them for any other violation." Sounds great, doesn't it? If that is the case, however, look for *documented proof.* In most cases you will not find it; it just is not being done. Whether management wants to admit it or not, they are in the "Linkage Mentality." In many cases, management thinks that safety rule violations are too difficult to prove. Let's look at documentation, a way to prove what we say.

## Policy and Documentation

What are your policies on safety rule violations? The best people to ask are your First Line Supervisors. Ask them to give you the company policy on tardiness, absenteeism, theft, and insubordination. Then and only then ask them, "What is the company policy on safety rule violations?" Do not ask them what they would do, but rather, what is the policy. Usually, they will try to compose an answer for your acceptance. If they cannot give you a policy on tardiness, etc., then you can be sure they don't know about a safety policy.

Let's go over some basic steps for safety policy enforcement:

1. If you do not have a policy, get one. Either write it, have Human Resources write it, or write it together. You cannot enforce a policy that does not exist or is so obscure that no one understands it.

2. Train your First Line Supervisors in its use. Either you and/or the Human Resource Director must go over it line-by-line explaining how the policy works. Explain to them why it was written, what is expected of them, and when it should be implemented.

3. The key to the success of this policy is documentation. If the Line Supervisor did not write it down, he did not do it! Wherever I go in construction and industry, I always ask about safety rule violation. I always get the response that, "We have a policy and we enforce it." But when I ask to see the policy or enforcement documentation, they cannot produce it.

Often, when I ask to see the documentation for safety policy they say, "I gave a verbal warning." Verbal warnings are an escape hatch for both the Line Supervisor and the erring employee. Do not allow verbal warnings! They are not remembered by either the Line Supervisor or the employee. They are also nearly impossible to prove at a disciplinary hearing. If the employee violated a safety rule that he knew he was supposed to follow, document it. Show this documentation to him, and discuss it with him. He will remember it better that way.

Line Supervisors like verbal warnings because no paperwork is involved. You will get the response, "All I would do all day is write warnings." That's the whole idea. The best way to assure that supervisors are doing their jobs is to get them involved. The best way to get them involved is have them assume the responsibility of writing warnings. The other assumed response is that if forced to write warnings, they will not write anything, claiming they did not see *any* violations. The best measuring device on this is the incident rate, your daily survey and the OSHA log. How can they claim they did not see an employee violate a safety rule when they have direct supervision of the employees all shift long? If this is true, we have a case of a Line Supervisor not doing his or her job. Either way, somebody, either the employee or the Line Supervisor, is not following safety rule procedure.

4.  I suggest the following course for documentation of employee written warnings. Please note: Each case must be documented on its own circumstances and severity. Do not fall into the trap of "it's our policy and we have to follow it." Write a policy that allows you to leap frog from one step to another.

5.  *Step One:* Write a warning regarding the violation of a company safety rule that will remain in the employee personnel file for one year.

    *Step Two:* Write a warning resulting in a suspension without pay for one-day for violation of the same rule occurring within a one-year period.

    *Step Three:* Write a warning resulting in a week's suspension without pay for the third violation of the same safety rule occurring within a one-year period.

    *Step Four:* Termination of employment for a violation of the same rule occurring within a one-year period.

    *Step Five:* Suspend from work for one week for a total of four written warnings regardless of the safety rule violation.

    *Step Six:* Termination of employment for the accumulation of five safety rule violations regardless of type.

Before you run screaming into the night, saying, "My company will never accept this!" please hear what Human Resource Directors say about it. I presented this scenario at a human resource conference recently. Half the participants were in shock that I would even suggest something so oppressive and harsh. The other half thought it was workable, acceptable, and efficient. The half who thought it workable had very few injuries, while the ones who felt it too severe had many safety problems.

# MEMORANDUM

**DATE:**

**TO:**     Company Employees

**FROM:**     Company President

**SUBJECT:** Disciplinary Policy on Safety Rule Violations.

We recognize that the most important resource of our company is our employees. We also recognize that we have a legal and moral responsibility to provide for those employees a workplace free from recognized hazards. Our responsibility to you also includes providing safety training so that you can protect yourself from injury while at work. Our safety training is given so that you can be knowledgeable about the only two things that cause accidents: unsafe acts and unsafe conditions.

We take your protection seriously and we insist that you take it seriously as well. As such, we have developed a disciplinary procedure for safety rule violation. The following steps have been created so that not only will you follow our safety rules and procedures, but also so that others will follow them as well and not put you at risk. Our safety rules are clear, concise, and easy to follow. If you need clarification, please see your immediate supervisor or the company Safety Director.

We will no longer be issuing verbal warnings. Any and all safety rule violations will result in written reprimands. These written reprimands will be reviewed by our Safety Director and Human Resources Director for technical accuracy and fairness. I do not foresee the issuance of any reprimands if all employees agree to follow our company safety rules.

These rules will be enforced predicated upon the severity of the incident. If, in the opinion of the company, a safety rule violation was so severe as to cause death or serious injury, then the employee can receive an automatic suspension or termination of employment.

This policy was written to protect you from the unsafe acts and conditions created by others.

**Disciplinary Rule #1.** You will receive a written disciplinary reprimand for violation of a company safety rule to include committing an unsafe act or creating an unsafe condition. It will remain in your file for one year from time of issue. If you do not receive another written reprimand within a one-year period, the reprimand will be removed from your records.

**Disciplinary Rule #2.** If you receive a second disciplinary reprimand for committing an unsafe act or creating an unsafe condition, within a one-year period for the violation of the same safety rule, you will be suspended without pay for one day.

**Disciplinary Rule #3.** If you receive a third disciplinary reprimand for the violation of the same safety rule, to include committing an unsafe act or creating an unsafe condition, you will be suspended for one week without pay.

**Disciplinary Rule #4.** If you receive a fourth disciplinary reprimand for the violation of the same rule, your employment will be terminated.

**Disciplinary Rule #5.** If you receive a total of four disciplinary reprimands within a one-year period, but not necessarily for the violation of the same safety rule, you will be suspended for one week without pay.

**Disciplinary Rule #6.** If you receive a total of five disciplinary reprimands for any safety rule violations your employment will be terminated.

Can your company take a stand this firm? Can you? *Safety really is a people business.* If you put the word out that safety rule violations will not be permitted, they will indeed cease. If you put out the word to your supervisors that they are going to be held accountable for the actions of their employees, they will, in turn, be accountable themselves. I know this is a tough road, but it is a very necessary step for you to do your job.

Wherever I go I am asked, "What is the one thing we can do to reduce injuries?" My response always is "training and the accountability and consequences for that training." Disciplinary action in the safety process of a company is an integral part of training that is often ignored. Your company will be looking to you for the unpleasant job of assuring that discipline is used in reducing injuries.

How is your safety committee getting along? Any problems? Does everyone shows up? Are all items are being addressed? In the next chapter, let's take a look at the misunderstood world of the safety committee.

# 8

# "They Won't Even Come for Free Doughnuts."
# The Safety Director and the Safety Committee

If there is one thing we keep hearing in our seminars, it is the ongoing complaint about the dreaded monthly safety committee meeting. In many cases, Safety Directors in industry regard safety committees the same way Safety Directors in construction regard hard hats—as something that's done simply because it's a rule. I am in favor of hard hats on construction sites just as I am in favor of safety committees in industrial settings. I do feel, however, that if an employee asks for a reason for the meetings, he or she should be given a better answer than "because it is a rule."

To begin with, a safety committee is not required by an OSHA standard. It's not even required for a successful safety program. It is, however, an excellent means to educate and involve your employees in the safety program. Please don't get the impression that you can't run a successful program without a safety committee—that is simply not true.

In earlier chapters, we discussed "hubcap safety," the program that makes your safety effort look good without really contributing to it. In many cases, I find that safety committees are being used for the same sort of dressing up.

I visited a company that had all the bells and whistles of a safety program in place. They had more paperwork, procedures, policies, addendums, and documented methods than the federal OSHA. Their Safety Director had recently been given this job and was clearly going to be a success. But all was not as it appeared. I was asked to do an audit to find out why people were still being hurt, even though the company was giving the Safety Director all the support she requested. When I went through my program checklist I was very impressed. I couldn't believe a company could have this much in place without maintaining a successful program. I was given reams of documentation for every meeting, agenda, and topic. I asked how she handled the line employee rotation. She said, "Well, you know that our committee is formed strictly on a voluntary basis." I acknowledged that I was aware of that. When I again asked about the line employee rotation she said, "We very seldom have members from our plant floor attend. We do, however, have middle management participate on a regular basis." The result was that no line employees ever attended the meetings more than once. Even though the meeting was on company time, the employees preferred to continue work rather than sit through, as one employee put it, "The longest two hours of my life."

Their safety committee was in place only because they believed a successful safety program was supposed to have one, not because they saw it as a tool to assist the overall safety effort. Think about that for a minute. What is your safety committee doing for the real core issue—the elimination and reduction of injuries in your workplace? Why don't you list what your safety committee has done in the past six months specifically to reduce the number of injuries? In order to do this honestly, you must track from the committee to the action. Can you do it?

*Safety is a people business*, and people are creatures of habit. Your committee may look great on paper, but if you can't prove what it has accomplished to protect your employees, then you're just getting together once a month to chat with your co-workers.

When I discuss this in seminars, our participants say that they totally agree with me, but their boss wants a safety committee in place—it just looks good to have one. "If the line employee doesn't want to attend, all the better. The company won't have to pander to their requests and we can run the committee the way we want to."

So your boss says you must set up a safety committee. Then why not have an effective one that really does the job of reducing or eliminating workplace injuries?

With all this being said, let's now go through the "What'" and the "Why" of a safety committee.

## What Is a Safety and Health Committee?

A Safety and Health Committee is a group that aids and advises *both management and employees* on matters of safety and health pertaining to plant or company operations.

## Keys to a Successful and Effective Safety Committee

### The Original Purpose of the Committee

What do you want—a committee or a team? A committee can be large and should act as advisor to the Safety Director. A team, on the other hand, should have a maximum of 10 members, including the Safety Director, and should act as a decision making group for the company safety program. Either a committee or a team can work to achieve the safety objectives.

Have you decided the purpose of your safety committee? In other words, what is it they have come together *to do*? The purpose must be clear and meaningful to the members of the committee. They must feel that the time spent in the meeting is significant and valuable to themselves and the company. The purpose should be defined in a written statement. This statement can be prepared by you, the Safety Director, or it can be developed by the safety committee as one of its first activities. Once the purpose is stated in writing it should be very visible to the committee members. From there on out, everything the committee does should be referenced in terms of "Does this fit our purpose?"

## Its Staffing and Structure

Who do you want to serve on your safety committee? Again, think of your purpose. A joint employee and management committee is strongly recommended. Employees can communicate problems to management, allowing information and suggestions to flow both ways. The committee can serve as a forum for discussing changes in standards, programs, processes, and potential new hazards. This approach can produce effective solutions to safety problems more easily. Because a joint committee facilitates communication and cooperation, it usually raises morale and commitment to a safe working environment and accident reduction.

Who from management do you want on your committee? Who has the potential to affect your safety efforts in money and decision making issues? I would suggest including the company President/CEO, the Company Comptroller, the Human Resource Director, an employee from maintenance, the Operations Director, and, of course, your boss, a Line Supervisor, and a number of line employees. Remember how important those Line Supervisors are to the success of your program.

When you have Line Supervisors and employee representatives on your safety and health committee try to have them stay with the committee for at least a year term. If these members change frequently, they never feel part of the committee and its important work. They will just be getting the hang of it soon before it's time to leave.

Structure your committee meeting with a written agenda, prepared and distributed ahead of time to the members. This will not only keep your meeting on track, it will also give members time to be prepared for agenda items.

Here is an example of a safety committee agenda:

---

**ABC Company**

**Safety Committee Meeting**

Date:_____

I.      Attendance

II.     Old Business - Reports on Assignments

III.    New Business

    A.      Near Miss Reports

    B.      Accident Reports

    C.      Safety Suggestions

    D.      Workers' Compensation Update

    E.      Monthly Injury Report Card

IV.     Open Discussion

V.      Who Does What, When

VI.     Conclusion

VII.    Next Meeting

---

Follow the meeting with written plans to follow up and complete the recommendations of the committee.

Here is an example of an action plan to follow up on items from the safety committee.

---

**Safety Committee**

**Follow-up Action Plan**

**Date:**_____

| Action Item | Person(s) Responsible | Resources Needed | Completion Date | Recommendation |
|---|---|---|---|---|
| 1. | | | | |
| 2. | | | | |
| 3. | | | | |
| 4. | | | | |

---

# Support It Receives While Carrying Out Its Responsibilities

Support from top management for the safety committee is absolutely necessary. Does management allow the committee to meet during the normal workday? Are members rewarded for their attendance? Nothing will kill participation faster than a boss punishing an employee for being off the line or out of the office. Is there resource (time, money, people) support for the work of the committee, both for the committee itself and follow-up on recommendations of the committee?

## Active Participation and Cooperation

Does each member, regardless of his or her job title, feel valued and comfortable speaking up at the meeting? If your CEO or someone from top management has a tendency to take over your committee meeting and make all the decisions, you may need to do some work on the side to get this person to understand your goal is to have everyone participate. Encourage all attendees to share their safety concerns in one minute or less at the end of the meeting. No judgment or resolution is allowed during this time. It is just a listening time. You can decide later whether to follow-up on the concerns or schedule follow-ups on the next agenda for group consideration and input. Continually reinforce that every member is an equal member of this committee. Over time, trust will increase and members will communicate more openly and work more cooperatively.

# Responsibilities of the Safety and Health Committee

Tasks of the safety committee can be grouped into four categories.

## Monitoring Tasks

- Recommending hazard elimination, reduction, or control measures.

- Planning improvements to existing safety rules, procedures, and regulations.

- Reviewing and updating existing work practices and hazard controls.

- Monitoring effectiveness of safety and health recommendations.

## Educational Tasks

- Participating in instructional programs.

- Compiling and distributing safety and health communications to employees.

- Recommending safety training.

- Evaluating the effectiveness of instructional programs.

## Investigative Tasks

- Inspecting the facility for unsafe conditions and practices, and also for hazardous materials and environmental factors.

- Investigating workplace accidents/incidents.

## Evaluative Tasks

- Assessing the implications of changes in work tasks, operations, and processes.

- Field testing personal protective equipment.

- Evaluating effectiveness of safety and health recommendations.

- Studying and analyzing accident and injury data.

- Studying and analyzing incident rate data.

# Running a Constructive Safety and Health Meeting

1. *Develop a clear mission or purpose statement and stick to it.* Make sure everyone has a copy of the mission or purpose statement and has read it. Refer to the statement often. Only deal with issues that fit your purpose. For example, if a personnel issue comes up, insist that the member take it to the proper department. In one safety committee, the issue of a radio being played loudly was discussed in the committee for three months in a row. It was not an unsafe condition but rather an issue of employees disagreeing on the choice of station and volume. This is clearly a supervisor and/or human resource problem. It should not have been discussed in the safety committee, much less for three consecutive months. Don't let your safety committee get distracted!

2. *Plan the details of the meeting carefully and write an agenda.* Preview visual aids and read handouts for accuracy and clarity. Have enough copies ready. Be sure audiovisual equipment is in good working order. Give members at least one-week notice of the meeting date, time, and place.

3. *Keep the meeting short and to the point.* Safety meetings should never last more than one hour. Meetings should usually be limited to one per month. If you have long and frequent meetings, they will become less effective.

4. *Keep the meeting respectful and friendly.* Let committee members know their participation is important. Encourage questions and ideas.

5. *Don't argue a point.* If a critical or negative statement is made, treat it is as just another statement. If someone appears upset or angry, acknowledge his or her feelings. Restate meeting objectives in a positive manner. Be nonjudgemental. There are no right or wrong questions.

6. *Be honest.* If there isn't an immediate answer or direction to take on an issue, tell the committee that it will be looked into and then get back to them with an answer as soon as possible.

7. *Always solicit suggestions for corrective actions on all issues from members of the committee.*

8. *Recap the meeting.* Summarize what was said to assure accuracy and reinforce important ideas.

9. *Give members the last few minutes to ask questions or state concerns.* End the meeting with a sincere "thank you."

10. *Follow up.* Do the research and investigative work necessary to answer member questions. Get an answer or information back to participating members as soon a possible. Share what you learned at the next meeting.

Most importantly, stick to your purpose. Restate it often. If the meeting begins to stray from the purpose, bring it back quickly in a friendly, respectful manner.

Safety committees do not have to be large cumbersome bureaucracies. Rather they exist to assist you and the company in reducing or eliminating injuries to employees and protecting the assets of the company. Use the meeting for that purpose.

Company politics has discouraged many company Safety Directors. Let's examine that subject in the next chapter.

# 9

# "But He's the Boss's Son."

# The Safety Director and Company Politics

Are you thinking to yourself, "Not only does this guy have no idea how bad my company's politics are, but he hasn't a clue on how to fix them!"? If you are indeed thinking this, that's perfectly all right with me. I am not going to try to impress you with horror stories from my career about the barricades of company politics, or nightmarish family held companies. No matter what I attempted to tell you, you'd probably say to yourself, "If you think that's bad, you ought to be here!" So let's start this chapter with the premise that your company's politics are unbearable, and they are getting in the way of protecting your employees.

The first question you should ask is "What specifically is the problem?" Whenever company politics comes up in a seminar, the response is laughter, rolling of the eyes, and "You do not want to know." If company politics is an issue that is getting in your way, then you have to deal with it. Safety Directors often throw up their hands and surrender. Internal political problems can be resolved successfully, and they are resolved successfully every day, but it takes a concerted effort and some political savvy on your part to do it.

## The Company Oracle

Okay, let's look at what specifically is giving you problems. I usually hear first about the "good old boy" syndrome. This commonly occurs with someone, usually a male, who has been with the company for a significant amount of time and who is considered immune from any real accountability. This person is a "hands off" individual. Every company has one, whether they admit it or not. He is a tough nut to crack because when they asked you to be the Safety Director, they never thought you would have the temerity to question "The Oracle." If, however, you are to do your job properly, you must question and even challenge anyone who is blocking your efforts to get the job done. The Oracle can be difficult because many people, including your boss, probably feel that the company simply could not run without him. This is nonsense—the company could run quite well without him, you, your boss, or most others. It's your job to be as professional and diplomatic as you can be. Remember, *talk straight and play fair*.

Just what is it that this individual is *either doing* or *not doing* that is obstructing what you want to accomplish? Is he duplicitous, unconcerned, belligerent, rebellious, or does he just find you and your position amusing? Does he talk out of both sides of his mouth, stab you in the back in front of the employees, or is he just ignorant of how important it is for you to have backing? Do you need his support? If not, then what you really want him to do is shut up or go away. I know it's tough talk, but it can be a very tough situation.

The first thing you need to do is list specifically what it is that he is doing to cause you problems. Please note: I mean *specifically, precisely, accurately, and exactly.* What behavior do you want him to change? What do you want him to do differently, or stop altogether? You cannot discuss these issues with him or your boss until you are clear in your mind what it is you want. You will need to list on a sheet of paper the consequences of his actions that are affecting the safety program.

Now is the time to be realistic. Are his actions really hurting your program or just your feelings? Believe me, I am not discounting how you feel about the situation, but you need to be completely honest with yourself. If this person genuinely feels that a Safety Director and a safety program are totally worthless for your company, then he is completely within his rights to feel that way. Oracles usually have been employed with the company from the very beginning and have seen it grow without a safety program. They are usually older and less aware of the need to protect people. They feel that common sense will protect anyone as long as it is used. You are not going to change that kind of mindset overnight, or maybe not at all. If this person sees you as the former secretary, assistant warehouse supervisor, etc., he will have a difficult time accepting you in a different position with an authority that impacts him. He may even resent it! Again, be very sure that he is hurting your *program*, not just your *feelings*.

Here is a measuring device we use in seminars to help participants to develop solutions for dealing with the Oracle and like species.

# MEMORANDUM

## Behavior

**Action/Inaction:** Does not follow personal protective equipment rules.

**Event:** On 11-23-97 he was observed running a grinding wheel without eye protection. When questioned by one of the supervisors he said, "It wouldn't dare hurt me."

**Consequence:** Employees in immediate area find the incident amusing. Since the Oracle has such respect within the company, this negatively affects the wearing of required PPE.

**Correction:** The Oracle should follow all PPE rules to protect himself and set an example for employees.

## Behavior

**Action/Inaction:** Is disrespectful to the Safety Director and the company safety program.

**Event:** On 2-17-98 he confronted me while I was discussing safety issues with employees. He interrupted me in front of the employees, and asked, "When is this crap going to be done? I have orders to fill."

**Consequence:** The employees became disinterested in our safety talk and the momentum and motivation of the meeting was lost.

**Correction:** During safety meetings the employee should wait until conclusion of the safety meeting or stay completely away. If he attends, he should make supportive and positive comments to set an example.

Does this look like a tattletale sheet to you? A way to inform on your fellow employees? Not at all. It is a tool for you to measure and record the actions of others who are getting in the way of your program. You need to do something that will chronicle your frustration and complaints. How are you going to look organized, professional, and competent when discussing this issue with your boss if you do not have yourself organized and prepared?

I can list hundreds of examples from my career and from seminars. I think, however, you get the idea of how to prepare a "Behavior Report." It can be used for a variety of issues, but it works well on this type of "people problem." Don't diminish yourself on this issue. You must take action to stop it!

Try sitting down with the individual and explaining the how and why of what he is doing that is harmful to your efforts. If you get promises made that are not kept, or you are dismissed as too sensitive, then the next step is to see your boss. The usual response in our seminars is, "You have got to be kidding. My boss would never say or do anything to the Oracle. He (she) is afraid of the guy." "He (she) will just tell me to do the best that I can." My question to you is "Have you done the best that you can with this person?" If so, state what you have done. List all the actions you have taken. Then review all those steps you have taken to make peace. Now, isn't it time to get some help from the boss? If not him or her, then who? If not now, when? Suffering in silence will only make you resentful and not solve the problem. Your job is to solve problems before they become a death or injury to your employees.

Another response from our participants is, "I simply do not have the time to do all this recording and recordkeeping to convince my boss there is a problem." "I have a job to do, I can't spend hours jotting down every single thing I don't like." Okay, I understand that. If this is a real problem for you, what are you going to do to resolve it? Sitting in a seminar or reading a book and nodding your head about how bad things are will not fix the problem. At this point, all you are doing is playing, "Ain't it awful?" Many Safety Directors play this game. They don't want to take the initiative to fix the problem; they just want to grumble about it. If there is someone in your company who is undercutting you and attempting to obstruct your efforts, you then must make the time to do what is necessary to end it. When people tell me they don't have the time, it's the same as saying a pilot does not have time to put fuel in the plane he or she is about to fly!

If you have done all you can do with the Oracle, and you are absolutely sure it's not your issue, but his, now is the time to see your boss. Send him or her a memo regarding the reason for your visit and an outline of what it is you need to discuss.

We had a woman in one of our seminars who had recently been appointed as the Safety Director. She had her own "Oracle" in her company, who also was a close friend of the company President. She stood in the seminar with tears in her eyes and said, "If I go to my boss, I'll be cutting my throat. I won't get any support from [the Oracle], the President, or my boss. I might as well just tell them to get someone else to do the job." She stood there by herself and then someone in the group starting laughing and then another person and pretty soon she was laughing. The point everyone got was that she could not do the job anyway, that's what was causing her frustration. When she left that day she was determined to put an end to the frustration, discouragement, and hopelessness she had been working with. She also understood that regardless of the outcome, she would have taken some action, and others would understand how she felt. So many Safety Directors believe that if they hold others accountable there will be severe repercussions. There very well may be, but is it any worse than enduring lack of support and ridicule? Aren't you already suffering repercussions?

Okay, so now its time for you to meet with your boss and the Oracle. Do not go in there to prove someone wrong, do not go in defensive, and do not go in cowed. Go to the meeting with facts, facts, and more facts. State why this behavior and resistance is a problem, why it must stop, whom it affects, and what needs to be done to correct the situation. Be professional and state what needs to be stated. Use the term "our safety program"—to emphasize that the safety program is suffering, not you. Emphasize also that the protection of our employees is being compromised because of this behavior, not that it is an issue because you are being ridiculed. Use what I call the Three P's: Be Prepared, Poised, and Professional.

## The Boss's Son

Remember when I asked you what support you were getting? One of the things you need to be aware of is that if there are family members involved, they will attempt to use their last name to have the final say. They, too, can cause problems like the Oracle. Determine from your boss how he or she is going to handle situations like this. Your boss is responsible for dealing with the family issues to assure that you can get your work done. It's not your job! Unlike the Oracle, who is a company employee, the family owners must understand that in order to provide a safety program, other family members must correct the errant one. I have seen this occur more than once and, almost always, it takes a family member to correct another.

What if your boss will not deliver the message to the family that some of them are getting in the way? If your boss lacks the courage to do the job, then you are the only logical choice.

What if he or she won't let you take such action? What if you get the "just do the best you can," song? In this case, you need to explain the facts

**119**

to him or her: that you simply cannot provide a quality program without resolving the problem. What is the worse that can happen to you? I always like to ask this question in seminars, because everybody stands up and yells, "I could lose my job!" I am always suspicious of Safety Directors who say, "Of course I want to protect my employees, but..." Either you have made a professional commitment or you haven't. Of course you could lose your job, but you could also lose your job for not properly doing your job due to the interference you have to deal with. Reread the ten things you do not want to hear from your boss.

## Running Safety Programs Through HR

Historically in American business, particularly in small businesses under two hundred and fifty employees, the Safety Director fell into one of two categories: "Pops Weaver" or the "Personnel Lady." Sound familiar? Pops Weaver was the fellow who for years did a fine job as a widget winder, but who was only two or three years from retirement and was suddenly in charge of "that safety stuff." Overall, Pops did a good job. He was not trained in safety engineering, but he had a good grasp on how an employee could get hurt or even killed in the plant.

As OSHA standards, industrial hygiene, and the numerous other requirements of safety evolved, companies began to turn to the human resources professional. The thinking in the front office was, "Safety really is a personnel function, so why not give it to Personnel?" This approach has been very popular for the last twenty years. The human resources department can fill out the OSHA log and take injured employees to the hospital, etc. Much to the surprise of management, however, the personnel staff began to walk out on the plant floor to look around. After attending a few safety seminars, they began to point out to supervisors ways to eliminate unsafe acts and unsafe conditions. This was a very interesting time for everyone involved. HR wanted to stop filling out OSHA logs and do the real "safety stuff." Supervisors wanted "them girls off the plant floor." The front office just wanted injuries to stop with an absolute minimum of disruption to production and quality. The result was a generation of in-house personnel staff who took safety seriously. Supervisors were not happy. They did, however, learn to get over it.

## Women as Safety Directors

Let me be just as clear as I can be on this subject: *Gender is not, will not, and never has been an issue on how well a safety program can succeed.* Gender, however, continues to be an issue when we discuss safety being a people business.

Historically, women were seldom on the plant floor doing safety inspections because they were either in the front office doing the typing or on the plant floor working with other women. Today, however, many women are managing excellent safety programs. Why, then, do so many women come to our seminars with such frustration? Most often because they want to do a good job but are hindered in some way from doing so. They care, but there is usually some man who is obstructing their efforts or being patronizing to them. Sometimes other women, too, can obstruct them, but in most cases it is a man, at least according to our participants.

Another difficulty for women in the safety field is the "Who, me?" syndrome that occurs in so many companies at the management level. If you are a woman, you probably know what I am talking about. Your boss, the President, the Superintendent, and many more all swear allegiance to "gender fairness" in the workplace. Behind closed doors, however, the conversation usually takes on the air of, "What are we going to do with her?" Women who are professional Safety Directors should be respected, listened to, and appreciated for their work.

Let me tell you a true story about an event in my career.

I cannot use her real name, so Betty Smith will have to do. As a Corporate Risk Management Director with three plants in three states, I needed a knowledgeable and trustworthy safety supply vendor. I had recently stopped using one after I caught him lying to me. A safety colleague recommended Betty. His comments were, "She knows more about our business than anyone I have ever met." During the interview, I asked her some perfunctory questions regarding safety to determine if she was knowledgeable about the business and about what some of my specific needs were. She responded favorably and I then hit her with my favorite question. I told her of an extremely hazardous chemical in my foaming operation and asked if she could get a respirator for my employees' use during that operation. She smiled at me and then made perfect eye contact. "You know and I know that there are no such respirators available. Only supplied air will protect your employees. I know my job, sir, and I know it better than any man or woman you will meet in this city or anywhere else. I have been doing this work for nine years and I learned in my first week that if I were to make a living as a safety sales representative, I would have to be faster, smarter, and better than another sales representative. I work harder, study more, and give better service than any other sales representative in the country. Give me a try and see what I can do."

I just sat there, dumbfounded. I honestly did not know what to say. She smiled and said, "The person who recommended me for this job told me that you could not be conned and you were a plain-spoken person. I hope I haven't been too plain spoken." I remember my words to her as if they were yesterday, "Let's walk around the plant and I will show you our operation." I am so very glad that I made that choice. She was everything my friend claimed her to be—she was perhaps the most knowledgeable person in safety equipment that I have ever met. She also taught me a lot about women in the workplace. Most often, they have to perform better than men to be treated equally. I do not like it—I have a daughter entering the workplace this year and resent the hidden barricades she will have to jump. But this is the same story I tell recently graduated female safety engineering trainees during their six-month training at a large insurance company. In the safety engineering field, you can be treated with regard, dignity, and respect if you *know* what it is you are talking about and can prove it. But, yes, it can be more difficult for a woman.

Now, here is the real rub. If you spent four or six years getting a safety engineering degree and had the barricades in front of you, you would still probably be motivated to do your job because you chose it. But, what about those of you who, for whatever reasons, have the job only because it was forced on you? Is this the time you march into the boss' office and say "enough"? That really depends, doesn't it? If you are interested and motivated, then wait and see how things are going before you jump ship. If you are looking for a way out, this is an excellent reason to do so. No matter what decision you make, be sure to include the employees covered by the safety program; they deserve that consideration.

## Speed Bump Politics

Company politics are speed bumps in the path to accomplishing what you need to do. Here is a story where speed bumps and politics literally met.

I have a client on the west coast with more than a thousand employees. The company has grown amazingly fast in the past two years and their facilities have expanded as well. With that many employees, the parking lot had expanded to accommodate all the cars. The parking lot was now huge with all the inherent problems of plant parking lots: speeding, reckless driving, inattention, groups of people talking and not watching, the usual problems you would find with this type of exposure.

It was becoming apparent to the safety committee that the parking lot situation would soon result in an injury or fatality. When asked, I told them to install a speed limit of 11 mph. "Why 11 mph?" They asked. "People don't drive 11 mph, they drive 10 or 15, but not 11." 11 mph, I said, required them to be focused on their speed. It was my intention to have them take their foot off the accelerator and let the car idle through the parking lot. Most cars will idle at 10 mph.

The idea worked well for awhile, until employees discovered there was no one monitoring their speed. They soon started to ignore the speed limit. No one in management wanted to assign anyone to watch the parking lot, particularly the company President. He felt his people had better things to do. When asked again, I suggested "one way in, one way out" that would automatically restrict traffic speed and flow. The President was against this because the traffic pattern would back-up traffic at peak hours at the main thoroughfare. It was then suggested and adopted by the committee that the only solution was speed bumps. They could not be ignored, would be effective, and get everyone's attention immediately.

The President was not at the meeting when this suggestion was adopted. At the next meeting, he vetoed it as being too restrictive for employees and overall an unnecessary nuisance. The committee tried to convince him that this was a good idea, but he would not budge on his decision. I know the President well. I know him for being a hardheaded pragmatist who really cares about the safety of his workers. I kept quiet on the subject because I knew the speed bumps were not going to be allowed and we were wasting our time. I also knew that when the opportunity presented itself, I would be able to discuss this with him.

Eventually the opportunity did present itself. The company had recently awarded lower pay raises than the year before. That issue and a few others convinced the President that there was no need to do anything else that would raise the ire of the already somewhat disgruntled employees. Speed bumps are indeed an irritant and why add more fuel to the fire?

Was he right in vetoing the speed bumps? Well, if I was in his position, I would have done the same thing. Yes, the parking lot was becoming dangerous, and, yes, an injury or fatality was possible. There are, however, political issues that can and do stand in the way of safety. If the issue was an imminent danger, I would have gone to the wall, but in this case and with this President's pro-safety track record, I let it drop. Eventually, the issue will be raised again and I will then attempt to initiate the dreaded speed bumps one more time. Remember, discernment. I think this is a good time to go into another politically related issue, the philosophy of prevention.

## Prevention Philosophy

I always tell seminar participants within the first few minutes that one of the best things that can happen to a safety program is a death or amputation. I'm sure your response is much the same as theirs: "How can you say such a thing?" I can say it very easily, because it's the truth! As brutal and insensitive as it sounds, it really is the truth. Let's use an example from the world outside the plant.

Two or three times a year, the local news will show parents marching with picket signs at an intersection they feel is unsafe for their children. If it's a slow news day, the local TV station will send out a crew to interview a mother or father to get a ten-second sound bite. We may watch this segment for a brief time and then see what the weather is going to be for the next day. Very seldom do we ever hear about the "Elm Avenue Safe Streets Association" again. That is, we do not hear about them again until the TV news crew interviews the grieving parents of a child who was run down at the same intersection. Then we see the footage of the first protest, followed by somber-voiced TV sages, including local politicians, who feel that "something should be done immediately to protect our children." This can all be put very simply into one time-worn phrase, "We don't shut the gate until the horse is gone."

Why does it take an amputation or fatality to wake up management to do the things you want them to do? Why does it take city hall to see the death of a child to get them motivated? While I cannot speak for city hall, I will share my thoughts on business management.

The safety profession is predicated on the "sky is falling" complex. We are almost like palm readers or astrologers in that we have to "predict" the future for our companies. Certainly we can use probability analysis and other tools at our disposal, but when we put ourselves in the role of the boss, we are always saying the sky could fall at any time. The sad part about it is that this talk becomes boring. We don't intentionally act that way, but that is the way we are perceived. No matter how much we want to change the way things are, we must accept the fact that there are times when we won't be able to do it. We need to accept that our profession is significant and valuable, but not everybody feels the same way, particularly our bosses. We can become roaring pains in the neck to them, and they would rather not get an adrenaline rush every time we walk into their office. The reason we have so much trouble with politics is because of our lack of discernment.

The next chapter on Workers' Compensation should really brighten your day!

# "They're All Thieves and Liars."
# The Safety Director and
# Workers' Compensation

Certain words elicit an immediate and disapproving response: murder, taxes, child abuse. For Safety Directors, those two words are "workers' compensation." Nothing raises the blood pressure and lowers the tolerance level more quickly than the utterance of those two words.

A Safety Director's lot is not an easy one, what with trying to get a decent budget, getting support from First Line Supervisors, convincing top management that occupational safety is more than a necessary nuisance, and trying to convince employees that you really care about them. Day after weary day of slogging through a job no one recognizes or appreciates. At the end of the day what is your reward? Some dishonest employee coming into your office with a questionable injury ready to ravage your incident rate and triple your days lost. Why, it's enough to make anyone contemplate dental hygiene school!

As one seminar participant so aptly put it, "I work and work to create a safe environment with no injuries and these thieves and liars can invent any story, tell any lie, and I have to accept it at face value." What is a

Safety Director to do? We are underpaid, overworked, underappreciated, and overlooked. Okay, now that we have that out of our system, let's take a fresh look at the subject.

## Elimination vs. Containment

As the Company Safety Director, you should be expected to stop workers' compensation fraud and abuse. You are *not* going to stop it, so you need to get pragmatic about your expectations. Your job is to prevent authentic occupational injuries and illnesses, not to be a police detective for people who do not play by the rules. Beating your chest and bawling to anyone who will listen will not do you or your position any good. I can not begin to tell you the number of very good Safety Directors who have burned out their careers dealing with this issue. If you plan on staying in this profession for any time at all, you are going to have to make peace with workers' compensation.

The first thing you need to understand is that it's not your job to stop workers' compensation fraud and abuse—it's the job of your workers' compensation insurance carrier. I worked for a national insurance carrier for over seven years as a Senior Risk Management Representative. I ran seminars for our workers' compensation policyholders on how they should use us to prevent fraud and abuse. I visited large accounts with serious claims problems and listened to how we, as the insurance carrier, would approve anything the employee claimed. Of course we would. The first report of injury would come in with no addendum from the client as to the particulars of the case and the claims representative would accept it at face value. Why wouldn't he or she?

Let me stop for a moment and ask you a question. Do you know the name of your insurance company claims representative? Do you know the name of your workers' compensation insurance carrier? How about the name of your insurance agent? If not, why not? These are the people you should be talking to about workers' compensation issues. You pay your insurance carrier a great deal of money for handling these claims for you. However, they cannot do the job for you if you do not give them the information they need.

Find out the name of your insurance agent and set up a meeting with him or her. Find out just what rights you have as a company to be involved with fighting fraudulent cases. In many cases, you will be told your rights to handle these cases has been given over to the insurance carrier. Yes and no. The insurance agent wants to keep you as a client and that means keeping you happy. Another way to keep you happy is to set up a meeting between you, your insurance agent, and your claims representative, and then demand that you have the same claims representative for your company. Regardless of what you will be told, it can be done if you insist on it.

Start looking at each questionable case and submit a letter to both your agent and the claims representative about your company's knowledge of the alleged accident/incident. Get statements from anyone involved in or with knowledge of the incident. Properly train your First Line Supervisors in accident investigation. Educate your employees about workers' compensation fraud and abuse and the cost to your company.

Write a policy on the consequences of filing a fraudulent claim at both the company and the state levels. All states have some type of legal consequence for fraud and abuse of workers' compensation. Talk to your agent and claims representative about the use of surveillance on questionable cases. All employees must be held accountable for their actions.

# You're the Guy with the White Hat

One of the problems with getting so involved with workers' compensation is that, before too long, you begin to believe that you and your company can make the final decision about the acceptance or denial of a claim. That kind of thinking can get you into big trouble. When we discussed elimination vs. containment, it still involved the overall responsibility for denial of a claim with your insurance company. You and your company do not make those kind of decisions and it's not up to you.

*Safety is a people business*, which means that you must maintain some type of isolation from the sticky wicket of being the guy in the black hat who says to your employee "Your claim is not believed and we are denying it." You do not want to be in that position, nor do you want to put your company in that position. Once your employees believe that you are taking an active role in seeing to it that their claims are denied, your safety program is dead.

The decision rests with the doctor and the insurance company. Your role is to gather the facts, not to have the final say on what the outcome will be.

Let me tell you a brief story about one of the dumbest things I ever did regarding workers' compensation. An employee came into my office claiming a back injury. The employee had reported it to his supervisor and the supervisor did the best he could with the facts he had. The employee couldn't remember where it happened, when it happened, or why it happened. All he knew was that the injury was "work related." I explained to the employee that there was no data to support his claim. The employee then angrily explained to me that the injury was work related and I was going to pay for it. It was then that I made an error that I have yet to forget—I tore up the report and threw him out of my office.

**130**

Within two hours, he was back with his union steward, of course, to tell me he now knew the who, what, when, where and why of his accident. The revised accident report read like the launch schedule for the space shuttle. The Union Steward informed me his fellow union member had a right to be confused with the original report due to his pain and suffering. By then I knew I had dug a deep enough hole for myself and was not going to attempt to argue the validity of the claim.

I called our insurance company claims representative and we discussed it over lunch. He came right to the point. "Next time keep your mouth shut, and let me do my job. Unless you can give me some uncontestable proof that this claim is fraudulent, I have to pay it." He was right, I was wrong, and the employee was inventive. I never made that mistake again.

From then on, I developed the Elephant Stampede Rule. An employee could claim that a herd of deranged elephants stampeded through the plant causing him great physical distress and I would accept it as fact on the accident report. It was up to my research and additional information to the insurance carrier to get it denied. It is not your job as the Safety Director, to argue the claim. You pay people for that purpose. Your job is difficult enough without going out of your way to make enemies.

## Workers' Compensation—Finger Tip Facts

The one common denominator among Safety Directors who get angry about the unfairness of the system is that they really do not know very much about it. Let's take a look at the system so we can understand what it is we are dealing with.

Workers' Compensation Insurance is required in almost all states, so that employees who are injured on the job with either a work-related injury or illness will receive medical care and compensatory wages until

**131**

they reach Maximum Medical Improvement (MMI). Most employers must secure either a state funded or private insurance carrier for workers' compensation coverage. Not all employees are covered by workers' compensation coverage. In most states, domestic workers, agricultural employees, and casual workers are exempt.

An injury is workers' compensation related if it "occurs in and out of the course of employment." At times, this definition can get a little murky. This is an area where good Safety Directors can help their program, their company, and themselves. Sit down with your agent or your claims representative and get a clear picture of what "in and out of the course of employment" means to you and your company. Who is the resident expert, or at least the technical advisor, to your company on this subject?

There are usually four types of disabilities for your employees:

1. **Temporary Partial Disability**–The injured employee will eventually be able to return to his or her routine job. He or she will either need time off, or modified duty, until they reach MMI.

2. **Temporary Total Disability**–Indicates that your employee cannot work, but will ultimately return to his or her job with either full or partial recovery.

3. **Permanent Partial Disability**–In this situation, the employee is permanently and partially disabled in some area of his or her physical ability as a consequence of an occupational illness.

4. **Permanent Total Disability**–In this case, the employee simply cannot return to work. He or she is permanently and totally disabled.

Your employees are eligible to receive rehabilitation. Usually it falls into two areas: medical, where the focus is on the medical recovery of the employee; or vocational, which covers job training and education when your employee will not be able to return to his or her original job with your company.

For all intents and purposes, workers' compensation is an "exclusive remedy" for injuries and illnesses arising in and out of the course of employment. In other words, your employee cannot sue the company because of injuries that occurred there. In return for not being able to sue, the employer must provide workers' compensation coverage for medical and rehabilitation recovery from injuries suffered while an employee.

In actuality, it's not a bad deal for both parties. As litigious as our society is, can you imagine the number of lawsuits that would be filed by disgruntled workers who wanted to take their bosses to court? It simply would end business as we know it today. However, there are also bosses and companies that take advantage of that system. The "If you get hurt, you can't sue me" mentality is alive and well throughout the country. So, exclusive remedy is telling employees that this is the only course of action they have. There are, however, as with everything else, exceptions to this rule. Let's take a look.

1. **Bringing Suit against Fellow Employees**–Your employees can sue co-workers, (including Safety Directors) who they allege caused them to be injured. They can claim that an employee was negligent in the performance of his or her duties, thus causing an injury. This occurs in an OSHA willful violation citation. Talk to your insurance agent and/or your claims representative about the laws in your jurisdiction.

2. **Bringing Suit against Another Contractor**–If your company is doing contractual work and another contractor's employee injured your employee, suit could be brought.

3. **Bringing Suit against Equipment Manufacturers**–This is probably the most popular exclusive remedy used. The employee sues the equipment manufacturer on whose equipment they were

injured. The manufacturer then sues the employer for not properly protecting the employee from the injury. Isn't safety fun?

## The Big Red Button—How Do You Know They Are Hurt?

Wouldn't it be great if each human being in your company had a big red button in the middle of his or her forehead, so that when a legitimate injury ocurred, or the employee was in pain from a nondefinable injury, you could see the red light flashing?

A perfect place to start would be with the dreaded "back injury." Ask any Safety Director in the business and he or she will tell you that back injuries cost him the most money, most lost time injury cases, and the most grief from the other employees. Back injuries, in many cases, are very subjective and they are hard to prove or deny. Oh, for the red button.

In dealing with workers' compensation, the best advice is to bring the injured worker back when you can find a job for him or her. Any job, just bring him or her back. That concept can save the company hundreds of thousands of dollars. It can also dramatically reduce your insurance modification factor and your incident rate. There is, however, some real grief associated with simply bringing the injured worker back to work.

## The Return To Work Process in the Real World

Read any book, attend any seminar, listen to any insurance agent or claims representative and they will tell you, "Bring the injured worker back to work ASAP." They are absolutely right, but be sure you know what you are in for. Remember, the name of this book is "Safety Is A People Business." Nowhere, at any time, will you need to use your people skills more than when you inform an employee at home to, "Come to work tomorrow, we have work for you."

Let's say you have done your homework and you know what it's costing you to leave your workers at home. You know that you have work available and you want to get the employee off the workers' compensation dole so you can reduce your already staggering premiums.

Before you pick up the phone, be sure you know whom you will have to deal with. In this area, the following people will have great input into whether or not you are successful: the Plant Manager, First Line Supervisors, line employees, and the injured employee.

## Plant Managers

In order to understand just what it is you want when you decide to bring the injured employee back to work, you must first put yourself in the place of the Plant Manager. Pretend for a minute that you are the Plant Manager. Your role and goal are to produce a quality product on time. As long as you do that, life appears to be pretty good. Now we add the Safety Director into the mix.

As the Plant Manager, you have met with the company President and the Safety Director and it seems that everything is going along okay. Sure, you have to make some inspections from time to time, and, yes, you have to get on a Line Supervisor about this safety stuff, but it seems to be working just fine. Then, out of nowhere, you hear that this new Safety Director wants to start bringing injured employees back to work.

If you're the Plant Manager, the first thing on your mind is "Where are we going to put them? We don't have any departments for the walking wounded, and if we put them on the plant floor, we have to watch them even more closely than we do our healthy employees." Also—and this is going to be a big problem—"When my supervisors get an injured employee back to work, they will start using this as an excuse for slow production and poor quality. If the problem will not be production or quality, then other employees are going to grouse about how they have to do all the work and we'll have nothing but labor problems."

This is an absolute no-win situation for the Plant Manager. If you were the Plant Manager, what would you do? Well, you would probably do the same thing he or she will do at the first opportunity: sit down with the company President or someone in high authority and tell them the reasons this return-to-work nonsense won't work! Return-to-work is great for the bottom line and great for the Safety Director, but a nightmare for the people who have to enforce it. So how do you convince the Plant Manager? By convincing your boss, or his boss, or whomever it takes, of the importance of bringing injured employees back to work.

As I said before, money talks, so talk money. Do your homework, get your facts straight, and be prepared to answer questions. If possible, have your insurance agent and claims representative in the meeting to assist and support your position. Discuss this issue with your boss in front of the Plant Manager, so that every time he or she raises an objection, you can counter it with financial savings. Make no mistake about it, this will be a tough sell, but a necessary one. If you can get the Plant Manager's commitment in front of the big boss, you can then go to your next obstacle—the First Line Supervisor.

# MEMORANDUM

**Date:**

**To:**        (The Big Boss)

**From:**     (You)

**Subject:**  Return-to-Work Program

1. As we have discussed during the past several months, our workers' compensation costs are increasing. Although the number of injuries that are presently occurring has leveled off, our costs associated with employees receiving compensation is still driving our high premiums.

2. At the present time, we have three employees off work receiving 66.66 percent of their pay from workers' compensation. It is becoming increasingly clear that we must implement a return-to-work program for those employees that can do some type of work, no matter how menial.

3. I have met and discussed this subject at length with our insurance agent and insurance company claims representative. They are adamant that the only solution for us is to start returning our injured employees to work. Other companies are using this method and their costs associated with their workers' compensation have dropped dramatically.

4. Considering my conversations with our agent and claims representative, we can reduce our expenditures with these three employees, who are presently home on full compensation, around 68 percent. That is the type of savings we need to reduce our insurance premiums.

5. I would like to meet with you to go over these figures so you can see the need for this program. I would like to have the Plant Manager at the meeting also, as these employees are (his/hers) and (he/she) needs to be informed on the subject.

Thank You,

(Your Name)

# First Line Supervisors

First Line Supervisors are the ones in this process who have the power to make or break the return-to-work program. They have absolutely nothing to gain in this program—nothing!

It represents nothing but work, frustration, and the fear that sooner or later someone is going to reprimand them for failing to follow all the rules. From their perspective, an employee of theirs is injured. They have already had to conduct an accident investigation, and explain to you and their boss why the injury occurred. Their incident rates and possibly pay raises and/or bonuses have been affected because of this. Now, when all the paperwork is done, the explaining is finished, and the assurance that this type of injury will not occur again in their department is over, you tell them the cause of all this grief, which they realistically couldn't prevent in the first place, is returning to their department to continue to make their lives miserable. As I said, there is nothing in it for them!

The whole purpose of the return-to-work program is to bring injured employees back into the workforce. Considering the employee's physical restrictions, the employee may or may not go back to his original department. If he or she can do some or part of the work, he or she may very well wind up there. If not, the returning employee may end up in another department. Now you will have two frustrated supervisors—the first line supervisor, who is already smarting from the original injury, and the other supervisor, who does not want someone else's problems. It can get very messy, very quickly.

So how do we handle this without a full-scale riot? Education, that's how. Before you attempt to start a return-to-work program you should educate your First Line Supervisors on the following:

**Why:** Why is it necessary for the company to initiate a return-to-work program? Emphasize the cost savings to the Line Supervisors. They are accustomed to hearing about money and it won't be a surprise to them. Depending on the climate and culture of the company, you can stress how it will benefit everyone. Be cautious of this approach. Supervisors who have been around for a while have the poker face of a Las Vegas card shark. When you start to give them the "win one for the Gipper" speech, you will never really know if they are laughing to themselves. When you mention money, be as specific as you can. Compare the employee staying home versus coming to work and the subsequent cost savings.

**What:** What will this program mean as far as work and inconvenience for them? Whether they ask it or not, that is what they are wondering. What kind of work will it mean for them? Do not give them some glib, patronizing answer that is not relevant to their world. Do your homework and find out how much of a mess it's going to cause for everybody involved. Tell them specifically what is going to happen in regards to their department when the injured employee returns. Tell them what they have to do to accommodate the employee's modified duty. Tell them what they should be telling their employees about the company's reasoning on this issue. Be honest. Tell them what it will entail on their part to accommodate the employee, but also what it will require to keep the other employees from causing personnel problems.

**Who:** Who will be affected by this program? Under what situations would an employee be eligible to come back on modified duty? This is a big issue for Line Supervisors. In a return-to-work program, you are not going to be able to bring everyone back, regardless of what the medical restrictions are. The goal is to try to accommodate as many injured employees as possible. There are going to be employees who simply will not be able to do the job regardless of how hard you try to navigate them. In many cases, this will be due to the type of injury, in others it will be due to their return-to-work attitude. Theoretically, you should be able to accommodate any employee to any job. Realistically, it's not going to happen.

Another "who" in this equation, is "who" is going to have the final word on whether or not the employee is going to be brought back? Is it going to be a blanket policy that everyone, whenever possible, is going to be returned? If so, do the Line Supervisors of the respective departments get a say in the matter? If not, why not?

Another very big issue concerns who will be counted as a full employee under the return-to-work rule? Many supervisors must meet a quota of product with the employees that they have. When an injured employee is brought back to work, they will want to know if he or she can be counted as a full-time able-bodied employee, or not counted at all. Many times Safety Directors and senior management forget about this issue, but the Line Supervisor does not. I always recommend that a modified duty employee be counted on the quota roster, if he or she is counted, as a half a person for recordkeeping. It always makes it more agreeable for the department supervisor. Be very considerate of the First Line Supervisor's feelings in this matter. They truly are the most important players in the return-to-work drama. Let's now take a look at the injured employees' co-workers.

## Line Employees

Now you have convinced your boss, the Plant Manager, and the First Line Supervisors of the benefit of returning injured employees back to work whenever possible. It's logical, reasonable, and bottom line effective. Do you think that argument will impress the other line employees? Not on your life!

In a perfect world, the scenario should go something like this: "Did you hear that the company is going to find work for all injured employees? No matter what the disability is, this company cares about us and will provide something meaningful and fulfilling for us to do to build our self-confidence. So now, rather than going home from the doctor's office with a no-work or light-duty slip and lay on the couch all day, I get to go back to work with my bandages and sling so I can feel like I still belong." Now, let's put ourselves realistically in the shoes of the line employee.

What do you think the line employee's biggest concern is when it comes to the return-to-work program? Unless you deal with or have dealt with the issue, you will be surprised. The absolute biggest complaint and concern of the line employee about a return-to-work program is fairness. Fairness in that they will not have to do more because the other guy is hurt. Fairness in that this injured person will not cause problems because of quota. Fairness—and this is the big one—in that the injured employee is really injured and not taking advantage of the company and the other employees.

I have operated a return-to-work program for my company and have started one for countless other companies. Every time I tell management about the fairness issue they always tell me that "It won't be that way in our company." But it always is. Line employees are deeply suspicious of being taken advantage of by the other line employees. A RTW program will do just that if you do not handle the situation carefully, sensitively, and professionally.

Here is a case that will prove my point. I will call the injured employee Skipper Thomas. Skipper had a minor low back strain and was off for two weeks before we were allowed to bring him back to modified duty in his department. All the medical paperwork was done perfectly, all supervisors were properly briefed on what Skipper could do, and all employees in the department were told what Skipper could and could not be expected to perform. When Skipper came to work, he was put on a job he liked and everything worked well. After a week, he was put on a job he didn't like and Skipper felt pain. Skipper was immediately taken to the doctor and his restrictions were not changed.

The doctor could not explain why this one particular job caused him trouble. The other employees, however, sure could. Skipper didn't like that job, there were also other jobs that Skipper didn't like and as soon as he was assigned there, he started hurting. When he was assigned to a job he liked, he stopped suffering. When asked why one job caused him pain and another didn't, Skipper would say, "How should I know? I'm not a doctor. It just hurts when I do certain jobs. If you don't want me to do the work that doesn't hurt me, send me home."

What do you think the attitude was of the other employees who were assigned the unpleasant jobs Skipper did not want to do? There was nearly an open revolt in the department. The Line Supervisor was ready to jump off a bridge with frustration, the Safety Director was made to look like an idiot by this employee, and the company overall gave the impression that they didn't care about the employees.

The reason the company started the RTW program was because it cared about its employees and needed them. The company knew that an employee off for less than six months will usually come back to work. Employees off work for more than six months and less than a year have a 50/50 chance of returning to work. An employee off more than one year will probably never return.

The employees were angry with the company for not caring about how they felt about the situation. The employees knew that Skipper was ripping the company off, but there was nothing they could do about it, except get mad at the company for allowing it to happen. The company also felt that Skipper was ripping them off, but there was nothing they could do about it. So everybody was mad or frustrated, except Skipper, who was fairly content.

In dealing with line employees, whether it's RTW or any other issue, remember the golden rule is *Nothing is real, unless it's personal.* The employees will talk as if they are one united labor front against greedy management, but, in reality, it's everybody for himself or herself. Line employees will follow your policies and procedures as long as they are fair. If they are not fair, they will create more problems than you can handle. They do not want to be stuck with solving the problems in their ranks caused by your high-minded ideas. Remember, fair is fair for everyone. If an employee is abusing your RTW program, you need to solve the problem one way or another.

In the case of Skipper Thomas, we simply had a meeting with the insurance agent, claims representative, and his doctor. We had the claims representative do a spreadsheet on what it was saving us. We then had to make up our minds whether the cost savings were worth the grief that he was causing the other employees and his Line Supervisor. We decided it wasn't. We then told Skipper and the other employees that he would have to rehabilitate at home, as the jobs we were giving him, although totally approved by his doctor, were aggravating his medical condition. Skipper went home and eventually got an attorney who tried to magnify Skipper's condition. Our insurance carrier finally offered a lump sum payment that miraculously cured all of Skipper's back problems. This was a case in which, no matter how good our intentions, it caused real problems with the other employees.

Your line employees will almost always know when another employee is for real or is using the system. Saving money with RTW is great, just remember not to lose your workforce in pursuit of cost savings.

## Injured Employees

Remember how the line employees felt about coming back to work after being injured? How they felt about being asked to perform some type of work even though they had been injured and were still in some discomfort or even pain? They probably felt it was just another way for the company to squeeze every bit of work out of them, regardless of whether they were hurt or not. It wasn't about keeping them focused on the job, or even about facilitating their recovery process, it was about saving money. In that regard, they are partially correct.

RTW is about saving your company money by bringing partially disabled workers back to work to reduce the staggering cost of workers' compensation.

To many employees, ordering an injured employee back to work is an unconscionable outrage. Once again, put yourself in their shoes. You did not ask to be hurt, the company is responsible for protecting you, the company failed and you suffered an injury. Instead of showing care, concern, and compassion for you, they send you off to some doctor, who minimizes your injuries, and says you can go back to work to do something.

Many of your co-workers are going to be looking at you as if you are trying to get a free ride, the same way you looked at others when they were hurt. Your Line Supervisor is going to be monitoring your every activity so you won't exceed your medical restrictions, and all of this because the company allowed you to be injured. You think you deserve time to rest and recover from this traumatizing ordeal in the peace, sanctity, and comfort of your own home. If you ever had any doubts about the way the company felt about its employees, you have no doubts now— you are an instrument to be used.

144

In many cases, that is the way injured employees feel. Now it's up to you, the Safety Director, to convince them, and their co-workers, that you want the same working partnership you had previously, regardless of their injuries.

I recommend conducting department meetings for explanation of the RTW program. Explain the purpose of RTW, how it will be used fairly and equitably. Explain that the company has no say whatsoever in whether or not an employee can come back to work—only the doctor can make that decision. Again, emphasize in the meeting the neutrality of the company in this critical decision making process.

Tell these employees that every RTW employee will have a medical restriction form that will be strictly followed at all times by the supervisor. Make sure to follow up that it really is strictly followed. Tell them that you wanted them as employees when they were 100 percent and you want them even when they are less than 100 percent. The company is not going to discard them if they cannot do all the work they previously could.

## "I Never Heard from You Guys."

Most companies like to give the impression that their employees are important. It's smart business to do that, and I highly recommend it. There are luncheons, Christmas parties, department meetings, and the company picnic. However, when employees are injured and cannot return to work, even on a part-time basis, they are often treated as if they fell off the planet. No one from the company calls them, sends them get well cards, or shows any interest in their overall well being. The mindset that, "the insurance company is handling it," often takes over.

Well, remember that the insurance company is *not* handling all of it. They may call occasionally to see how employees are doing for medical reasons. But to employees, this call is from a stranger whom they have never met, and who has only a professional business interest, not a personal interest, in them. If you want employees to come back on RTW with a positive attitude, the best way to start is to show some interest in them while they are still at home. Safety is, indeed, a people business.

As I have mentioned several times before, training is the single most important thing you can do for the employees and the safety program. Let's look at that subject in the next chapter.

# 11

# "You Just Can't Train These People."
# The Safety Director and Employee Training

If you must learn your profession, and you must, then think about what your employees must learn in order to work safely. OSHA has numerous mandated training requirements that you, as the employer, must provide for them. I hear constantly from Safety Directors, both new and experienced, about the unfairness of it all, how OSHA dictates their training expectations with no thought or regard as to how to implement the training. I also hear, en masse, how "these employees simply cannot be trained." As we discussed earlier, however, the most important thing you can do that will give you the experience and credentials with your employees is to provide the training they need.

What I often hear from our seminar participants is, "No matter what training we give them, they simply will not follow it, or they don't pay attention, or they find it all a joke." As one Safety Director said, "My employees say to me, "No matter what happens, it's the company's fault and I'm not responsible for any injury!" Is that happening to you? How is your training program coming along? Do you have a company training program? Do you wonder why injuries are up and compliance is down? Let's take a look at safety training as described by the Occupational Safety and Health Administration (OSHA).

OSHA has a set of voluntary guidelines for the employer. They were written to help employers to protect their employees from workplace injuries and illnesses. These guidelines are:

- To determine whether a workplace problem can be resolved by training

- To determine what training, if any, is needed

- To ascertain goals and objectives for the training

- To plan learning activities

- To manage training

- To determine the productiveness of the training

- To modify the training program based on feedback from employees.

The benefit of safety training for your employees is immeasurable. Quite simply, without this training your employees will not be aware of the hazards they confront every day. The better the recognition of hazards, the better the mindfulness of these hazards your employees will have. This mindfulness saves lives and reduces injuries.

## Where To Start

I promised I wouldn't turn this chapter into an epistle on the benefits of safety training. Just keep one thought in mind, *safety training of your employees is the single most important thing you can do!*

Where do we start? Why, at the beginning, of course. The beginning is OSHA-mandated training. There are two schools of Safety Directors on the subject of OSHA training: The "This is a waste of time" school, and the "This is needed, relevant, and critical to the safety of my employees" school. Care to guess which school I recommend? OSHA has many mandated training standards predicated on the exposure your company creates from its operation.

Ever wonder where OSHA came up with all these training requirements? Contrary to what many Safety Directors may believe, they are not dreamed up by elves at a factory in New Jersey. These training standards resulted from the injury and death of many American workers. These standards were bought with blood and lives. Because of that, these mandated training standards are very important to the health and safety of workers. Let's take a look at one that seems to drive everyone crazy: CFR 1910.1200 Hazard Communication.

For years, employees worked with the mindset that business owners would notify them of dangers associated with their work. An employee who was exposed to chemicals only had to ask the supervisor if the chemical to which he or she was exposed was hazardous and they would be told the truth. Regrettably, some companies, or at least supervisors, were somewhat stingy with the truth because they did not honestly know what the hazards were, or they did not want the employee to know of those hazards.

Picture a new employee working with an exotic chemical who begins to have skin irritation and difficulty breathing. He asks his supervisor whether this chemical might be the cause. The supervisor responds that there is nothing dangerous in the chemicals he is exposed to and to just keep working.

So where is the new employee supposed to go to find out the hazards associated with his or her work environment? Before the Hazard

**149**

Communication Standard, employees really did not have a source of information, at least at work. With the advent of HAZCOM they could, by law, immediately look at the Material Safety Data Sheet and see for themselves if the exposure was dangerous to their health. Not a bad system. I think even the supervisor would agree. Not all mandated training requirements can be as beneficial, but the best place to start is with the OSHA standards.

The first thing to do is find out what standards are relevant to your company's exposures. Get out your trusty CFR and go through the standards to determine if your company falls under the training requirements. Admittedly, the CFR can be difficult to research and there are other sources that immediately give you the training requirements. No matter how you determine this information, you must obtain it to get started. Remember that OSHA's federal and state plans have consultative services that can usually assist you with a phone call. *The very first thing you need to implement is mandated OSHA training.*

## How To Train

Our company conducts nearly all of the required OSHA training. We go coast-to-coast dealing with frustrated Safety Directors who are overwhelmed with the requirements. But that is not all they are frustrated with. They are also confused and overwhelmed with how to do the training and the type of training to use.

In our profession today, it seems that every week a new safety training video, booklet, interactive communication, or pamphlet is coming out on every subject. Our mail is littered daily with a new "quick fix" video or booklet that "will save time and money." They claim to turn disgruntled, malcontent employees into the safest workers in your plant. Companies request our support and endorsement for them. I never recommend these "smash and grab" products. They are like the neverending supply of diet wonder pills. Take two pills at night and the pounds are gone in the morning! These safety training items may or may not do the job for you. The majority of these safety training items are very expensive and usually only work as a one-shot attempt. Many Safety Directors do not have the budget for this slick, glossy glitz and they do not need them.

Hubcap safety also encompasses all the new safety training trinkets that are on the market. Using these items may make you feel good, but it does not ensure that your employees will either learn or apply the information on the job. Isn't your point in training to assure that your employees apply what they learn? Undigested information is like undigested food—it just sits there in a lump.

Videos are excellent and I use them a great deal. However, you do not need videos to keep your employees' attention in safety training. What will always get your employees' focus is personal accountability. Employees will listen to whatever you say if they absolutely know that they are going to be held *personally accountable* for what is being taught. That is the true key to safety training—the knowledge that they are personally accountable for their own training is the best training you can give them.

## Training Handouts

Let's use Hazard Communication training as an example of how to write an employee training handout.

## Sample Cover Sheet for Handouts

**Date:**

**Hazard Communication for Compliance with CFR 1910.1200**

**Employee Name:**

**Department:**

**Instructions:** This Hazard Communication Safety Training is mandated by Acme Industries and the Occupational Safety and Health Administration. This programmed learning will require you to fill in the blank with the required information when presented by your instructor. The instructor will state the answer and also point to it on the overhead screen. You are encouraged to ask the instructor for assistance or information at any time during this presentation. Upon completion of this presentation, you will be required to take a learning exercise to measure what you have learned during the presentation. You may use your handout material to assist you in this learning exercise. Please refrain from talking with other participants so that everyone can hear the instructor. Please remember that your safety training and its application of content are a condition of your employment.

1. _____are one of the first and best places to look for information on the materials you are using. (Answer: Warning labels.)

2. The two most common types of labels are_____ and _____. (Answer: Manufacturer's written labels and color-coded labels.)

3. There are four classes of chemicals. They are _____, _____, _____, and _____. (Answer: Flammable, corrosive, toxic, and reactive.)

4. Some examples of flammables include_____and _____. (Answer: Alcohols and gasoline).

I'm sure you get the idea. Overheads are a very good idea because you want to control the participants' eyes. You can make the overheads yourself from your handout material. Point to the answer and tell them the answer. The purpose of this type of training is to put the safety training accountability where it should be—with the employee. Using this method requires the employee to, first, *think* about the answer, thus engaging the brain; second, *write* down the answer, and once again think about the response. The important point here is that employees must think about what is being said. If they know they are being held accountable for what goes on in the training session they *will* be more accountable.

You can include as many items in your training handouts as you wish. I am, however, a proponent of training sessions less than forty-five minutes in duration. This is not a Madonna concert; it's safety training. No matter how fascinating you and I find the subject material, your employees probably will not.

## The Learning Exercise

Once again, the learning exercise is an excellent tool to hold your employees accountable for their training. The material for the learning exercise should come directly from the handout material and, if possible, be in sequential order. You are not trying to make any employees look foolish, but give them information they can apply on the plant floor. If you try to get tricky with their learning exercise they will know it and deeply resent it. What we want them to think and tell their peers is "If you pay attention and fill in your handout material, you will pass the learning exercise." Here is an example of a learning exercise from the training we just covered.

# HAZARD COMMUNICATION

## Learning Exercise

Name:

Date:

Department:

Please answer all questions of this learning exercise by filling in the blanks for each question. You may use your handout material to assist you. When you are finished, turn this learning exercise into the classroom instructor who will grade it and return it to you. If you need assistance, ask the classroom instructor.

1. _____ are one of the best places to get information on chemicals.

2. The two most common types of labels are _____ and _____.

3. Two of the four classes of chemicals are _____ and _____.

4. One example of flammables is _____.

You can make the learning exercise as difficult or as easy as you want. In this case, I just paralleled the handout material. Again, do not get tricky in this activity. The employees know you have the power to make them look stupid and they will not forget the way they are treated. It will harm them, you, and, worst of all, your Safety Program if you purposely make the exercise too difficult. Also remember that many of your employees have not been in a classroom in years. This is going to be a very threatening situation for them. If you simply tell them all they have to do is write down what you tell them, it will lower their anxiety level.

One quick point concerning training of your employees: I often see Safety Directors, both new and experienced, who want to impress their employees with their vast knowledge of the profession. They go out of their way to let everyone in the room know that they have read a few books. Believe me when I tell you this—the employees don't care! They already assume you have some knowledge or you would not be allowed to train them. I do not mean to stand in front of them and say, "I don't know any more than you, I was stuck with the job." Rather, just do your training and they will discover for themselves that you have some knowledge on the subject.

We recently conducted supervisor training for a large utility company on the subject of "Researching and Understanding OSHA Standards." Twenty-five whiskered, gnarly veterans waited for me to conduct the training. One of the participants asked if he had to call me "Doctor". It seems another Ph.D., from a local college had done some training for them recently and had insisted on the title. I told them of course not, "Mike," or "Hey, you," would be fine. My best advice is to get out of your ego and into educating your employees.

There are going to be times when you will have employees who read and write poorly or do not read and write at all. Prior to the training, ask their supervisors whom they suspect have this challenge. It is then up to the supervisor to assist this person outside of class in this learning. Also, check with Human Resources as to who might have difficulty in training. There are cases when the supervisor hasn't a clue as to which employees in his or her department need extra help with these skills.

## The Supervisor and Employee Training

Do not make the mistake that so many Safety Directors make regarding training of employees. Do not let the Line Supervisors use this as an opportunity to take an hour off! Line Supervisors must be actively involved. They are as much a part of the training of their employees as you are. Line Supervisors need to be in the room for at least one of the training sessions, and their assistants, if they have any, need to be present for the additional training. What we are talking about here is the *attentiveness and comportment* of the employees. We have talked many times in this book about the role of the Line Supervisors and their impact on the safety program. It is in this area of classroom training where they send the message to their employees that safety training is part of their job and they will be held accountable for it.

---

**at•ten•tive•ness** *adj*   Concentrating one's attention on something.

---

That "something" is the subject of your safety training. Please note: Attendance alone does not count! Simply showing up for mandated training and then slouching down in one's seat, zoning out, daydreaming, or just not paying attention, is not doing what the employee is being paid for. If the employees were on the plant floor, they would be expected to be attentive to their respective tasks. The same is true for classroom training. Your company does not simply rent their bodies for an eight-hour day. They hired the whole person, including their minds, to perform as so directed.  Employees agreed to that when they were first hired and they must be held accountable for it during their safety training. It is in this area where the Line Supervisor, not you, must assure their attentiveness. Be prepared for Line Supervisors to complain about your unrealistic expectations. I run into this situation a lot, so let's look at how it can work for you.

When I am called into a training situation, I usually like to find out who I will be training—are they plant employees, supervisors, managers, etc.? Then I need to know why this training is being conducted by me. Also, I need to know what set of circumstances has necessitated my doing the training as opposed to the company Safety Director or another supervisor or manager? I usually discover that there is a problem with the knowledge level of the employees and the company feels that I can do a better job of getting the message across. When I discuss training with the Safety Director, I usually get the story that, "These employees don't care and there isn't much that can be done with them." I always ask whether the Line Supervisor attends the training and assures the attentiveness of the employees. The Safety Director's standard response is, "The supervisor is too busy to attend the classes; he (or she) just sends them to me." If I were a supervisor, I would love that system. No responsibility, no accountability, and, best of all, no consequences if my employees don't pay attention.

I always tell the company Safety Director that I like to meet with Line Supervisors prior to my training. I tell these supervisors that the company has hired me to do safety training of their employees and ask whether there is anything I need to know about the group. Usually the supervisors say nothing of interest. I then tell them that I expect them to attend the class and be responsible for it. I tell them that I am paid to educate, not to baby sit. I tell them that if their employees are not attentive, I will point them out to the supervisor. It is the supervisors' responsibility to deal with the problem, not mine!

When Supervisors hear they must be in the room and hold their employees accountable, they cannot believe it. They say they are too busy, they have already had the training, they have a meeting, etc., etc., etc. The point is that I do not baby sit classrooms; I train adults. Supervisors are responsible for the safety of their employees. That includes going to the training and assuring their employees are attentive. You, as the Safety Director, must hold supervisors accountable for this activity.

---

**com•port•ment** *n.*    One's actions in general or on a particular occasion

---

You measure the comportment of your employees every hour of every day while they are at their workstation. Why not do the same when they are in safety training? A big reason some employees cannot be trained is because they are not being held accountable for the training, and/or their supervisor is not being held accountable for the training. I deal with classroom comportment in the same manner as classroom attentiveness— the Line Supervisors must deal with it immediately. The subject of workplace safety is too critical for us to shove employees into a seat, and shove a video into the VCR and hope for the best. That sort of training is being done every day and it is not working.

Supervisors are going to cry out, "What do you expect me to do?" My response is, "Do whatever it is you do on the plant floor when an employee doesn't do what he or she is paid to do." Your time as the Safety Director is too valuable to not hold all parties accountable for learning how to work safely.

If you have attended any of my seminars or training sessions, you know that I don't come across as a grump, like I might appear to be from my comments on these pages. I enjoy humor, give and take, and encourage as much feedback from the group as I can get. However, in order for the group to feel safe and comfortable in their learning process, there cannot be a diversion or lack of classroom etiquette. Remember, these people are adults, so treat them like adults. If, for whatever reason, they do not want to be there or learn, let a higher authority handle the problem. It is your job to educate, not discipline. That is why they have supervisors. Make sure the Line Supervisors do their job, so you can do yours!

## "The Lord Is All the Safety I Need"

Occasionally, you will run into an employee whose religious faith exceeds what you would deem prudent for personal safety. In my conversations with other Safety Directors, I have heard stories about "'How the Lord will protect my eyes, feet, back, hands, etc." I have no intention of questioning anyone's personal faith. In many cases, when I find myself running into a brick wall of noncompliance based on faith, I often just tell the employee that the company must follow the mandates of the OSHA law, and if in good conscience they cannot comply, we wish them well in their new endeavors. In many cases, that is all you can do.

Often, when I am conducting a training seminar and sense that one or two participants are getting ready to give me a quotation from the Good Book, I tell this joke.

There had been a heavy flood for the past week. Harvey, a local farmer, was standing on the front porch of his farmhouse with water lapping at his feet. A boat with two of his neighbors came by. One neighbor yelled, "Jump in the boat, Harvey, the river is rising and you will drown." "No," Harvey said, "The Lord will protect me from the flood." Three hours later, another boat came by and found Harvey hanging out the second story window of his house with the water level just under the windowsill. The boat passengers called out, "Harvey, get in the boat, you are going to be swept away." "No thanks," yelled Harvey, "The Lord will protect me from the flood." Four hours later, a Coast Guard helicopter flew over Harvey's house and saw him perched on the roof, with the water once again threatening to sweep him away. "Attention, you on the house," blared the loudspeaker from the helicopter. "Prepare to grab the rescue cable." "No thanks," cried Harvey, "The Lord will protect me from the flood." Two hours later, Harvey was swept away by the rushing water and drowned. When Harvey got to Heaven, he met St. Peter and asked for an audience with the Lord. The audience was granted and, upon meeting the Lord, Harvey said, "Lord, I have been a loyal and faithful servant of yours my whole life. Why didn't you protect me from the flood?" The Lord replied, "Harvey, I sent you two boats and a helicopter—what else did you want?"

Safety training requires doing all you reasonably can to help others help themselves. If an employee, for whatever reason, be it lack of attentiveness, poor conduct, or faith, fails to avail himself or herself of this knowledge and training, it is your responsibility to correct the situation by disciplinary action. You can train your employees, but sometimes not all of them. If they can't be trained then they shouldn't be allowed to be on the plant floor. Read and re-read the General Duty Clause. It lists specifically what you must do to protect employees. Within those words is the implicit mandate that your employees have to be trained.

Now let's now take a look at accident investigation in the next chapter.

# 12

# "It's Never Their Fault."
# The Safety Director and Accident Investigation

Safety is indeed a "people business" and one of the hardest tests for that concept is the accident investigation. For some reason I have yet to determine, it is in this area that Line Supervisors and Safety Directors often have the most difficulty. In our "Accident Investigation" seminars, we have a variety of exercises that require our participants to role play. I divide the room in half and take one half outside into the hallway. I assign them roles of injured employees to play. Some are belligerent and hostile, some are frightened, and some are attempting workers' compensation fraud. Others are open and honest and want to assist in the investigation. What our participants understand is that this is the way the vast majority of the workforce responds when they are interviewed, initially by their First Line Supervisor, and then by you, the Safety Director.

This chapter will not begin to cover all the critical aspects of the accident investigation. That would take a whole other book. I want this chapter to aid you in focusing on the "people" part of this important element of your safety program. It is the manner in which the Safety Director conducts the accident investigation that will help determine success or failure for the entire program. Please do not minimize the significance of the accident investigation.

There is no other one-on-one, face-to-face interaction in the safety program that has as much consequence for the Safety Director as the accident investigation. You can teach, train, and coach your employees about how you and the company care about them, but when it comes time to determine whether an unsafe act or unsafe condition caused the accident, that is when the employees will buy in or bow out.

I teach the basics of accident investigation using the standard formula that has been around for years: the Who, What, When, Where, How, and Why. I have seen just about every type of accident investigation form in existence. Some are one page long and some are up to fifteen pages. Some are being used for "hubcap safety" rather than for real causative reasons for the accident. Regardless of the style or length, if they answer the six questions of accident investigation, they will do the job.

As you go through these steps, please focus on the people part of the investigation. In other words, how do people relate to these questions? So many times, Safety Directors and Supervisors get so obsessed in "just the facts" that they fail to reflect on the injured employee. Again, *safety is a people business* and it is never more true than in accident investigations.

## The Six Questions of Accident Investigation

### Who

- Who was involved in the accident?

- Who was injured?

- Who witnessed the accident?

- Who reported the accident?

- Who notified emergency medical services personnel?

## What

- What happened?

- What company property was damaged?

- What evidence was found?

- What was done to secure the accident scene?

- What level of care did the injured employee require?

- What was being done at the time of the accident?

- What tools were being used?

- What was the employee told to do?

- What machines were involved?

- What operation was being performed?

- What instructions had been given?

- What precautions were necessary?

- What protective equipment should have been used?

- What did others do to contribute to the accident?

- What did witnesses see?

- What safety rules were violated?

- What safety rules were lacking?

- What new safety rules or procedures are needed?

# When

- When did the accident happen?
- When was it discovered?
- When was the accident reported?
- When did the employee begin the task?
- When were hazards pointed out to the employee?
- When did the supervisor last check the employee's progress?

# Where

- Where did the accident happen? Be specific.
- Where was the employee's supervisor when the accident occurred?
- Where were co-workers when the accident occurred?
- Where were witnesses when the accident occurred?
- Where does this condition exist elsewhere in the facility?
- Where is the evidence of this investigation going to be kept?

# How

- How did the accident happen?
- How was the accident discovered?
- How were employees injured?
- How was equipment damaged?
- How could this accident have been avoided?

- How could the supervisor have prevented this accident from happening?

- How could co-workers avoid similar accidents?

## Why

- Why did the accident happen?

- Why were employees injured?

- Why did the employee behave that way?

- Why wasn't protective equipment used?

- Why weren't specific instructions given to the employee?

- Why was the employee in that specific position or place?

- Why was the employee using that machine or those tools?

- Why didn't the employee check with the supervisor?

- Why wasn't the supervisor there at the time?

As you go through the information in this chapter, please refer back, again and again, to the six basic questions and relate them to the people you will be involved with.

## The Supervisor and the Accident Investigation

Please stop and think back for a moment on what you have learned in this book about First Line Supervisors. First, you may have to persuade, convince, or intimidate them into buying the safety program. Then you must insist they become responsible, accountable, and knowledgeable about the safety of their employees. You must also insist that they suffer the consequences for not properly doing their jobs. They should ensure that they, as well as their employees, will attend all your safety training and will learn and apply the knowledge.

If there is an accident, the Line Supervisors must figure out a way to bring the injured employee back to work in the department and keep the rest of the employees happy. They need to complete a professional accident report that you will check, re-check, and question them about.

Supervisors as a whole do not like to fill out accident reports. They do not like the paperwork involved, sitting face-to-face with their employees, or answering to you. In most cases, they will do whatever possible to avoid what they consider to be an unnecessary waste of time. Once again, you will be required to hold them accountable and once again they won't like it. It is your job to ensure that the way in which they conduct the initial accident investigation is done appropriately.

## "You're Making Me Look Bad"

If you set up an accountability system for your Line Supervisors, and you should, they are going to resent employees who get hurt. Often the face-to-face investigation and report writing will not be a pleasant experience for the injured employee. This is an area in which you must insist that your Line Supervisors are *fact finding* and *not fault finding.*

Many Line Supervisors will use the accident investigation to absolve themselves of any accountability in the injury. As was mentioned earlier in the book, no one wants to be held accountable. Supervisors are no exception. Line Supervisors will be delighted to bring you reports that are brimming with the sins of the employee, while the Line Supervisors themselves appear totally blameless for the injury.

The other issue you must be concerned with is the way in which the injured employee will view this experience. In many cases, the employee can only assume that the Line Supervisor is representing you, the company, and the safety program as a whole. Many of your line employees are economic prisoners, who need their jobs and will do just about anything to keep them. That makes them vulnerable to the coercion of the angry and resentful Line Supervisor. You must be aware of all these variables and potentialities. How do you handle it?

Training Line Supervisors in proper accident investigation is not difficult. There are a variety of excellent accident investigation courses available. What is not available is your ability to train the supervisor to talk straight and play fair. You must write and deliver that message yourself. Here are some techniques I recommend:

The most important rule, and your Line Supervisors won't like it, is that Line Supervisors must accompany the injured employee to the hospital. Nothing will show the company's care, concern, and commitment more than that. Line Supervisors are responsible for the safety of their employees and must demonstrate it by this simple act. They should be there to look after the employee, to call his or her family, to call the company, and to answer questions he or she may not be able to answer. The Line Supervisor is there as the representative of the company and the safety program. You will hear a million reasons why Line Supervisors are too busy for this task. Since when are Line Supervisors too busy to look after the employees they are charged with protecting?

First Line Supervisors should be the ones to complete accident reports. They know the employee better than anyone else in the company, and they must represent the company's care, concern, and commitment to this employee. No one else in the company has that visibility and responsibility.

First Line Supervisors must sit face-to-face with the employee and complete the accident investigation report. The Line Supervisor should ask the questions and have the employee answer them. It is here that Line Supervisors must look for facts, regardless of where the facts may lead them. The Line Supervisor must be fact finding, not fault finding. The injured employee must *know* that.

Line Supervisors must not question the injured employee before the employee has received adequate medical care and is returned to work. Questioning the employee on the way to the hospital or during medical treatment is not permissible. The message you want the injured employee to receive is, "Your injury is the most important thing to us; we will get the facts later."

The first question to ask an injured employee is about medical treatment received. Did he or she get immediate care and was it satisfactory for the injury? Place the emphasis on his or her care, rather than what may have been done wrong. Do not put employees on the defensive, but rather let them know that you care about them as human beings, not just as a pair of hands.

Begin the interview with, "We need your help to see that this type of accident never happens again." Make injured employees a part of the solution, not the problem. Tell them, "For some reason, we allowed you to be injured, and we want to correct that." For many, the word "allowed" is a tough pill to swallow.

---

**al•low**  *v.*   Neither to neither forbid nor prevent.

---

When your employees are under your control and they are injured, you have allowed it. The law and OSHA's standards regard the employer as having control and either allowing or prohibiting activities in the workplace. To attempt to word it another way with the injured employee would be pointless.

Always understand that you as the representative of the employer have control and are totally responsible for the prevention of occupational injuries and illnesses in your workplace, according to federal and state law. Beginning your interview with statements like, "What rule didn't you follow?" or "How did you get yourself injured?" will only put the employee on the defensive. Many Line Supervisors will attempt to do just that so they will not have to accept responsibility. Make sure you educate your Line Supervisors on the acceptable way to open up an accident investigation meeting.

Supervisors must complete the entire accident report. Do not allow them to skim through it answering only the questions they believe will bolster their assertion that there was no prevention possible on their part. When they do a poor job on the report, they are sending you a message: "I'm too busy for this nonsense. If you don't like it then you do it." Be very consistent and firm on this issue—the Line Supervisors must complete the form properly.

The accident report form will ask specific questions about how the injury occurred and what was done to prevent recurrence. This is the heart of the accident report. Contrary to what many Line Supervisors believe, the accident report is not a punishment for allowing an injury to an employee. It is a tool to ensure that this accident will not occur again. Do not allow Line Supervisors to brush over how to prevent the injury. Never allow them to put, "Be more careful." or other such answers on the report. Determine what conditions and actions occurred to allow the injury to happen.

The Line Supervisors must show professional facial affect during the interview process. If they are smirking, rolling their eyes, or refusing eye

**169**

contact, it will again make the employee defensive. Supervisors need to understand that often they are not in control—the injured employee is. In the workers' compensation process, the interviewer does not want employees creating scenarios of what happened that are at variance with the facts. Treat the injured employee with respect. Line Supervisors must also understand that any employee at any time can simply walk out the door to his doctor and attorney and claim a work-related injury. Regardless of what your state law says about the injured employee seeing your company doctor, employees can and will make your company regret it if they are not treated with respect.

The accident report form should be either printed or typed. Have your Line Supervisors use black ink (it reproduces much better than blue), and insist that you or anyone else should be able to read and understand it. They should read it back to the injured employee to assure that what is stated is what the injured employee actually said. If your company has a policy of having the employee sign the statement, then have it signed and dated by the injured employee.

Your accident investigation training program will give you the type of information you and your supervisors will need to complete the accident report. Safety is indeed a people business and the way an injured employee is treated during an accident investigation is critical to the safety program.

## "Now What Happened?"

The Safety Director has a very important role in the follow-up to the accident investigation. A Line Supervisor simply filling out the report and slapping it on your desk is not the end of the matter. It is with you that the real investigation of the injury begins. You control how the "people part" of the investigation gets handled. Please do not minimize this role in the process; it is crucial to your program. Let's go over the recommended techniques for Safety Directors to use in reviewing an accident report.

- Unless the injured employee is hospitalized or at home recovering, you should demand to receive the completed accident report within 24 hours. Meet with the Line Supervisor first and go over the report *line-by-line*. Use the six basic questions listed above as your guide. Ask penetrating questions regarding training (remember, the employee was not trained if it wasn't written down!), equipment, experience, etc.

- Get the Line Supervisor to offer an opinion of the employee's honesty in the report. In his or her opinion, what factors other than the incident itself may have contributed to the report—divorce, money problems, chemical abuse, dislike of the Line Supervisor? The more you can get the Line Supervisor to give you a candid appraisal, the more information you can use to determine the cause and contributing causes of the accident. This part of the accident investigation is as important as the Line Supervisor's role in interviewing the injured employee.

## "Where Did Our System Fail?"

After you meet with the employee's Line Supervisor, you then need to meet privately with the injured employee. These are some of the items to cover.

- The first words out of my mouth to an injured employee are, "How are you feeling?" Then I ask, "How was your medical care here in the plant and at the hospital?" I want to know how our Line Supervisors and our first aid program performed for this injured employee. Did he or she have to lie on the floor while people ran around confused and panicked, or were the assistants caring and professional? How was the care at the hospital or clinic? Did he or she have to wait for treatment or was the treatment immediate? Did the Line Supervisor accompany him or her to the medical care facility? How did everyone involved treat him or her? Does he or she have any suggestions, comments, or opinions on how the company can improve its medical treatment of employees? Begin to involve the injured employee in the accident investigation process. This employee knows you are going to be questioning him or her about the accident, but it is not necessary to be hit with questions immediately.

- The next thing I always ask the employee is "How did our system fail?" I then explain our company's responsibility to him or her, that by law we are responsible for protecting employees. I tell the injured employee that we had total direction over the injuring action and there must have been some areas we were deficient in. Was it a lack of knowledge, experience, or did we not enforce our policy on unsafe acts or unsafe conditions? Where did our system break down? I want the employee to get an overview of our intent. *I emphasize over and over again how the purpose of this meeting is to ensure that this type of injury never occurs to anyone else again.* I tell the employee that I need his or her help and simply cannot prevent another similar injury without it.

- Spend time going over the accident report with the employee. Go through it line by line so that he or she knows that you are interested in his or her views and opinions on the subject. Ask how they were treated during the interview with the Line Supervisor. Ask for suggestions on improving that process.

- After you have gone through the report in depth with the employee, ask the question, "What would you have done differently?" In most cases, the employee will give you an honest appraisal of how the injury could have been prevented. It is during this conversation that the question of unsafe act or unsafe condition will usually occur. It is usually during this interview that the employee will ask if he or she is going to "get into trouble" because of the accident.

- Always try to get all the opinions, facts, suggestions, and comments from the injured employee *before* asking what he or she would have done differently. It is usually then that the unsafe act or unsafe condition issue is raised. I always answer the discipline question by stating, "This is the investigative stage. I will meet with the Plant Manager and Human Resource Director to review the accident regarding disciplinary action. I want you to know, however, I am very impressed with your candor and suggestions to help prevent this from happening to another employee."

Whenever we go through the above process in our seminars, we usually have one or two participants who say, "It will never work in my company. Our employees won't open their mouth about anything to help prevent accidents. They will just sit and stare at me." My question always is "Are the questions you are asking the injured employee causing the problem, or is it the attitude the employee brings into the meeting?" They always respond that it is the attitude of the employee. When asked why the employees have such a negative attitude, their response is usually that it is because the employee doesn't like the company, the supervisor, etc. If this employee falls into the minus 10 percent we discussed earlier, there really is very little you can do about it. If, however, this employee reflects poor supervision from the plant floor, that is a whole separate issue you need to address, at least as far as your safety program is concerned. The first place to begin with is the Line Supervisors. With your help, they need to convince the line employees of the sincerity of the program. This can be done through training and through your Monday morning safety talks. If it is a safety issue, it must be addressed in the safety meetings. If it is another issue that affects your safety program, it must be addressed through whatever department will assist you in resolving it. Don't ignore the problem; do your best to correct it.

The next step is to review all the facts you have from the accident report, your interview with the Line Supervisor, and your interview with the injured employee. There are two issues you must address:

1. What can you do to ensure this accident will not happen again?

2. Was this injury caused by an unsafe act on the part of the employee?

In Item #1, you must carry the load initially to ensure that all steps are taken, and that all necessary people are notified on how the prevention of recurrence will take place. You need to meet with any and all parties involved to take whatever action is needed. Your accident investigation training will give the information and insight on how to prevent future recurrences from the situation. However, your training will not guarantee that you will take the necessary steps. It is up to you to do that. Safety programs often fail because Safety Directors, for whatever reason, are great gatherers but poor disseminators. *Get the word out to anyone and everyone involved on how the injury occurred, and what is to be done to prevent recurrence.*

Now it's time to address the disciplinary part of this accident. After reading the first twelve chapters of this book, you know how strongly I feel about unsafe acts. That being said, I also feel just as strongly about talking straight and playing fair. Review all the facts in the case. Did the employee commit an unsafe act? Why did it occur? Was he or she not properly trained? Did he or she have limited experience, limited equipment, or limited assistance? What role did supervision play in the accident?

Factor everything into this decision. If you have determined that there was an unsafe act committed, then list your facts point by point. Meet with the Plant Manager and Human Resource Director. They will usually give your opinion great latitude, because, after all, this is your area of expertise. State the facts professionally: the employee was trained; had help, knowledge, experience; knew he or she could have requested help, etc. But, in this case, the employee simply chose to ignore or forget the policy and procedure and committed an unsafe act. Also state if the unsafe act created an unsafe condition for other employees. Were any co-workers at risk because of his or her actions? Usually the Human Resource Director handles the disciplinary letter and process. Be very clear that you are following the rules and procedures as outlined by the company, and you are confident of your facts.

Of all the activities in your position, this is one of the most important. You are in the position of determining, based on your knowledge and experience, whether or not an employee was injured from an unsafe act or unsafe condition. You will be in the position to recommend disciplinary action to be taken against the employee. It's not an easy position to be in, yet you need to do it. Your employees will watch the way you handle yourself on these issues. If they determine that you do talk straight and play fair, you will succeed in this people part of safety. If you don't, you won't.

Spend time learning the basics of a good accident investigation—master all the elements involved. Become proficient technically in this area. With strong technical knowledge and good people skills, you will succeed in the field of accident investigation.

Let's now go to the core of safety being a people business—communication.

# 13

# "If Only They Would Listen to Me."
# The Safety Director and Communication

This book is about the "people part" of the safety profession. When you interact with people as the Safety Director you must be able to communicate effectively.

| | |
|---|---|
| **com•mu•ni•ca•tion** *n.* | Something (as information) conveyed by writing, speech, or signals. |

There are excellent publications on the subject of communication. This chapter will give you a brief summary of what tools you need to start with. As a Safety Director, hopefully it will whet your appetite to pursue the topic further. I strongly recommend you make communication a topic of continued research.

How well you communicate will determine how well you and your safety program will succeed. In the past, Safety Directors were usually good communicators. Pops Weaver, who was given the job because he was experienced and close to retirement, usually knew the language of the plant floor and could get his message across. But Safety Directors who are educated more as engineers than behaviorists are less able to communicate effectively with Line Supervisors and line employees.

Often the engineers know a subject well, but they cannot or will not take the time to explain it to the line employees. When I train and mentor new Safety Directors, I always tell them that, "Knowledge is explanation. You don't know it if you can't explain it." Safety Directors need to explain in plain English the how, what, and why of the safety program.

## Safety Communication Obstacles

- *Production noise.* Have you ever been on the plant floor when an employee approaches you to ask a question? What do you normally do? Most Safety Directors will stand there and yell back and forth with the employee, hoping he or she will understand what they have to say. Does a surgeon in an operating room have to yell instructions to his or her assistants? How about a commercial pilot—does he or she yell back and forth hoping instructions will be understood? Is your profession any less critical in the protection of people?

  Always strive to answer the employee's question in an environment where you can understand one another. It demonstrates respect for the employee and it strengthens your communication.

- *Language.* Your ability to communicate with employees is crucial to your program. The language you use in the boardroom is not the language you should be using with the employees on the plant floor. First, they probably won't understand you in those terms, and second, they will probably resent you. Be respectful of line employees in your communications. You may have differences in background, education, and job experience but on the plant floor you should use their language!

- *Planning your communication.* When you are conducting safety training, safety inspections, or accident investigations, do you plan for it? Why not do the same thing before you attempt to talk to an employee? I am not suggesting that you study and cram for hours prior to walking the plant floor. However, if there is a specific subject that you will be discussing with an employee or supervisor, then plan for it. Know what it is you are talking about. You are, after all, your company's technical expert. Again, *if you can't explain it, you don't know it*!

## Organized Communications

Are you in the position of achieving what is needed for the protection of people and property? This requires you to be systematic, organized, and efficient. In the area of communication, you need to be just as systematic, organized, and efficient. Your boss, the Line Supervisors, and the line employees need to understand accurately what is expected of them. This requires clarity and precision in your communication skills.

- *Make your requirements understandable.* In safety engineering, you must always factor in that few in your company will grasp the entirety of what it is you are talking about. When you discuss personal protective equipment, for example, you can picture the various types of equipment from head to foot. Your supervisors and employees, however, may only be able to picture safety glasses. Be very exact in what you want. Keep your instructions simple. Conceptual communication will fail you on the plant floor. Be literal in your requests and you will achieve more.

- *Always know your audience.* Telling a veteran supervisor what you want is very different from what you tell new employees during their orientation. One changeless safety fact is that employees will always appear to understand more than they really do. I suggest asking them questions to determine their comprehension. This should be done without arrogance or a condescending demeanor. One way or another you must make sure they understand your instructions. Quite often the way you say something, as well as what you say, will stop an injury or save a life.

- *Be professional.* Whether you were selected for the Safety Director's job, or ordered by your boss to do it, the responsibility is yours. I can think of very few positions in a company that require a higher degree of professionalism than yours. Leave your personal feelings out of your communication. Be clear, concise, and courteous in your directions. Be positive in your tone of voice and facial affect.

- *Know when to say what.* Every order or instruction you issue will not be a matter of immediate life or death. We discussed discernment earlier. It is just as important to know *when* to say something as it is *what* to say. If the situation concerns imminent danger, then the gloves come off and you get the message delivered. If it is a routine matter, then a request is often more acceptable to the supervisors and the employees than a command. Experience will assist you in this area.

- *Know safety policies and procedures.* You will be perceived as the expert in your company on safety policies and procedures. When I do a risk assessment in companies, I ask Safety Directors about the policies and procedures, and almost always their response is uncertain or unfamiliar. If you don't know the policies, then who does? Line employees will stop you on the plant floor to ask for clarification on certain points, especially regarding discipline. Line Supervisors can only be as effective as they are knowledgeable. If you want the support of your Line Supervisors, don't put them in the position of having to guess or translate your policies and procedures.

- *Follow the intent of the letter of policy.* Supervisors, employees, and human resources administrators will call on you to give clarification. Every policy and procedure was initially written with a purpose in mind. There will be times when you alone must be the one who interprets the policy. Always make this interpretation with the intent of providing the most protection for your employees. Unless you do this, you will be seen as a hindrance to production, employee morale, and the efficient operation of the production line. You, however, are the one who is being held ultimately responsible for the safety of the employees.

- *Introduce new rules, policies, and procedures correctly.* Changes in safety policies and procedures can be very difficult for supervisors and employees. Handled correctly, however, you can make the transition much smoother. The first thing you need to do is meet with the Line Supervisors and inform them of the who, what, when, where, why, and how. Ask for their questions, comments, observations, and suggestions on the subject. They may have insights you have missed. Next, you must publicize these changes to the employees. Use bulletin boards or department memos—give the employees time to become accustomed to the change.

- *Decide who is going to sign the memo.* I often go into businesses where all company correspondence to employees is signed "Management." Who is that? It always appears to me that someone in "management" does not want to shoulder the responsibility of his or her decision. Unless it is a violation of some company decree, please take responsibility for your own safety policies and procedural changes by signing your name. The employees will know anyway that you are responsible and it gives you more credibility with the people on the plant floor.

## You Can't Hear When You're Talking

Again and again in this book I have emphasized that you will be considered the company expert on safety. It may very well be that you are. That does not mean, however, that you know automatically more about the operation and its exposures than anyone else in the plant does. Your employees, Line Supervisors, and many others have a better grip on that subject. Employees often complain to me that, "The Safety Director never listens." There is no one position in a company that needs listening skills more than yours!

- *You must exercise good faith in your listening.* By "good faith" I mean you should trust what others have to say and want to listen to them. Like every other Safety Director, you will have in your company employees who will attempt to monopolize your time with pointless, unnecessary, time-wasting questions or comments. When you see them coming, you will want to run in the other direction. I wouldn't dare suggest to you that you spend hours listening to their chatter. However, if you listen well enough to at least understand their point, it will allow you to make the decision to hear more or excuse yourself for a more pressing matter. Please use good faith in listening.

- *Jumping to conclusions.* If there is one ailment that Safety Directors suffer from it is jumping to conclusions. You hear the same question so many times that your response becomes automatic. This can be a dangerous habit to get into because without really hearing the employee out you can give bad advice or direction. If you consider the question too simple or too complex, you may have the inclination to take a dangerous leap off Mt. Conclusion. Think your response through. Process the question and your response.

## "Do You Understand?"

Your attempt to determine if the other person really understands what you said is called feedback. This is important in the safety profession. The orders, directives, and instructions you give quite often will be the difference between life and death.

- *Make sure that you are always available, approachable, and accessible.* Let's go back to Pops Weaver for a moment. As you remember, Pops was given the Safety Director's position because he was close to retirement and knew the operation well. Pops was always available, approachable, and accessible. Why? Largely because Pops didn't have an office. Pop's office was the plant floor. If he needed to file paperwork, he did it in the cafeteria. You could always find Pops and he was always there to answer your questions. As the safety profession evolved, the Safety Director was given an office. Even worse, the office had a door. Even worse than that, the door was usually closed. The Safety Director then was given the authority and status in the company of middle management. So now Safety Directors sit in their offices, with the doors closed, and wonder why the troops don't visit. An excellent way to get feedback is to become available, approachable, and accessible.

- Want feedback from the employees on how they feel about your program? Here is an idea: *Ask them*! When you make your tour on the plant floor, ask the employees what they think. Ask for their opinions, suggestions, observations, and comments. You will be surprised at how much they have to offer. You will not get this feedback if you do not ask.

- If the only time your employees see you is when there is a problem, they will soon equate your presence to a negative role. Feedback is much easier to get when the employee feels that he or she is in a safe environment. If you are perceived as being a positive person and employees can approach you as you make your rounds, they will be more forthcoming with information. Give out a few "Great jobs!" for a job well done. It works.

## Safety Meetings

Next to root canals and tax audits, the safety meeting is usually one of the least popular activities of your employees. A meeting is any gathering for the express purpose of giving or receiving information regarding safety.

Safety meetings are usually attended by the following participants: Shoe gazers—the folks who are enamored by their shoes and who can't tear their eyes away from them; Star gazers—the employees who raise their eyes to the roof of the building looking for anything but safety information from you; Deer-in-the-headlights—the employees who have a plastic glaze to their faces as if they have been hit by the high beam of an oncoming car. There are many more variations, of course, but you get the idea. Let's look at some ideas that have worked for Safety Directors.

## Planning a Safety Meeting

- *Deciding on the Purpose.* Why are you having the meeting? What is the purpose of the meeting? What do you hope to accomplish? These are the absolute first questions you need to ask yourself. Be very specific about what you want to achieve.

- *Deciding on Attendance.* Who really needs to be at the meeting? Safety Directors can obtain a terrible reputation for wasting everyone's time if they require the attendance of those who don't need to attend. Don't earn that reputation. Have only the people you truly need at your meeting.

- *Allotting Time.* How much time should you allot for your presentation? How much time should you plan for group discussion? How much time is needed for questions and answers? How much total time is needed? Keep your meetings as short as possible. Remember the three Ps: Be poised, prepared, and professional. Don't waste other people's time; they'll resent it.

## "Okay, I Think We'll Get Started."

There are times when the only words in a meeting that make any sense are "Let's get started." Most people do not like the meetings they have to attend. Many feel slighted if they aren't invited, but they also don't like to make time to attend a meeting. Besides, the attitude might be that all Safety Directors are fanatical nuts who believe the sky is falling. Most department managers know how to look interested and sincere, while, at the same time they are praying that you will be succinct and to the point. Why not give everybody what they want?

- *Getting things started.* All literature on this subject will tell you to call the meeting to order on time. There are no special words for this purpose, other than what is comfortable for you. Your next sentence, however, must include the *purpose* of the meeting. Let your attendees know why they are there. This is very important. Be brief and to the point as to why they have to attend this meeting.

- *Priming the pump.* Give your attendees some background regarding the purpose for the meeting. The who, what, when, where, how, and why so they can have some direction in their thought process.

- *Blending and harmonizing comments.* Most meetings need a referee, for want of a better word. If the subject is hotly contested or appears to embark on someone else's turf, you will need to keep the comments on track. It is your responsibility to maintain a focus on the objective. Here, however, is the tough part. If this is your meeting, with your agenda, and your vested interest, you will have to be more objective than anyone else is. Remember the three P's. Be poised, prepared, and professional.

- *Summarizing.* It will be your job to conclude the meeting. Go around the room getting concluding remarks from everyone involved. They may not want to be there, but they do want to be heard. Has the meeting achieved its objective? If not, why not? What needs to be done to accomplish the objective? What do you need to do personally to see that it is done?

One last point about meetings. The reason many meetings fail to accomplish much is because they never determine *who does what, when?* I can picture the countless meetings I have attended where the only real accomplishment was the sipping of coffee and consumption of doughnuts. Always end your meetings with *who does what, when.*

# Dressed for Work

You're appearance says as much about you as your word. The advice I give Safety Directors is very simple—dress down. Let me give you an example of a young man I trained a few years ago who had a difficult time understanding his role and personal appearance.

He had been selected as the Assistant Director of Human Resources and the company Safety Director. His previous position had been in the front office as a Transportation Coordinator. He was well suited for the position with a willingness to learn and an interest in Human Resources and Safety. In his coordinator position, he would often wear short sleeve shirts or golf shirts. Middle and senior management staff were required to wear ties and jackets. After he was promoted, he immediately went out and purchased a new wardrobe. He came to work looking like a Philadelphia lawyer. There was nothing wrong with that, except he looked out of place walking the plant floor and attempting to conduct inspections in this wardrobe. The line employees, who had worked with him previously, thought he was becoming arrogant and condescending based only on his appearance. The Line Supervisors were concerned about his ability to represent safety while wearing a tie around moving machinery. I attempted to broach the subject with him, but he was adamant that he had to dress this way to comply with company policy. I met with the Human Resources Director and Plant Manager. They both agreed that it was not necessary for the Safety Director to meet a certain dress code, as his position was unique in the company. As Assistant Human Resources Director, he was also required to walk the plant floor. The solution was resolved very easily in this case. Now he is back to wearing golf shirts and casual pants and is doing a very good job.

I am a firm believer that an individual should dress for success, and that appearance makes the position. In our profession, for a variety of reasons, the key is to dress down, not up. Remember your audience.

I suggest you use what I call the *three E's* when dressing for work: *Employees, Equipment, and Exposures.*

Your ability to communicate is critical to the success of your program. No matter how knowledgeable you are, or how many certifications you have, if you cannot communicate your message, your program will fail. Do whatever it takes to increase your skills in this area. Remember, *if you can't explain it, you don't know it.* Good luck.

Let's now turn to the next chapter and discuss your personal commitment as the Safety Director.

# 14

# "It's Not That Big of a Deal."

# The Safety Director and Personal Commitment

I thought about putting this chapter first so you would be more motivated in going through the rest of the book. After careful consideration, however, I decided it would better serve the reader to go through the book and then make an informed decision about this chapter. By now, you know that I have a passion for this profession. I view it as one of the greatest gifts one person can give to another—the protection of another person's well being and safety. I sincerely believe that.

## Two Common Denominators

The quotation for this chapter title came from a young woman at a seminar, a Human Resource Specialist who was rather disgruntled that she would have to handle safety, too. She surprised some of our participants and found agreement with others. To her way of thinking, it really was not a big deal. She viewed the job as simply filling out some forms and telling people to be safe. In many cases, that is all Safety Directors do, right? Indeed, there is a whole population of Safety Directors out there who feel that way. They all have two things in common:

A. They are more worried about protecting their position, than the employees' safety and lives.

B. They don't know what they don't know.

## Nothing Is Real, Unless It's Personal

Whenever I train Safety Directors and Line Supervisors, I put them through the following exercise. It always has a strong impact on them and hopefully it will give you some insight.

I want you to picture someone you care about very deeply. Maybe it's your spouse, one of your children, a mother or father—anyone who is a very important part of your life. Now imagine that this person has had a life-altering work injury. Imagine that this individual was running a machine and had his or her dominant hand severed at the wrist. Imagine that the hand simply could not be saved and this person will have to go through life with prosthesis.

First, let's focus on the injury's effects. What activities will this person have to stop? Softball, sewing, swimming, bowling, skiing? How will it affect his or her interaction with the immediate family and friends? Will there be difficulty in cooking, cleaning house, tinkering with cars, doing yard work? Will his or her ability to dance be altered? How about his or her self-esteem? Will this person become reclusive and fearful venturing out in public?

Now, how would this injury affect you? First, you would get the call that this person had been injured and you will be told what hospital to go to. Then you will probably be told by the doctor or nurse of the extent of the injury. You will then see the person and offer comforting words about doing whatever is necessary to assist in this difficult time. As the shock of the injury wears off, and you understand that although the person will live, he or she will be physically changed for the rest of their life, you will start to become angry. You will start to ask yourself, and others, "How could this have happened?"

## "These Things Happen"

At this point, you would probably call the company where this person worked, and insist on meeting with the people involved to determine how this tragedy could have happened. Let's now go through a scenario of how the meeting might proceed.

The Company President and Human Resource Director might say, "We are deeply saddened by this terrible accident. Your____was a valued member of our company team and we are just sick about it. Is there anything we can do for you or your family?"

You: "No, thank you. ____is feeling better and the medical care is good. But, my family and I can't understand how this could have happened. You're supposed to have a company safety department that watches over these kinds of things, aren't you? What happened?"

Company President: "A complete investigation is underway. Also, OSHA has done an investigation and we have been cited for a serious violation. We will of, course, appeal this citation as we feel our company has done nothing wrong."

You: "But what happened? How could something like this occur?"

Company President: "Apparently, the machine was shut down for repairs. The maintenance department failed to replace the guard. Your_____ started using the machine without the guard in place. He (she) knew better than to do that, but did it anyway."

You: "Are you trying to say that it was his (her) fault?"

Company President: "It's really nobody's fault. As regrettable as they are, these things sometimes happen no matter how hard we try to prevent them."

You: "But, you said OSHA cited you for not guarding the machine. Doesn't that mean you were negligent in your responsibilities?"

Human Resource Director: "We can understand your frustration and anger at a time like this. But the company President and all members of our company feel we have done all that was humanly possible to prevent the injury. We are confident, and so are our attorneys, that we will be absolved of all culpability in the accident."

You: "But, he (she) has lost a dominant hand! He (she) can't drive the car to pick up the kids, play basketball, or go fishing. He (she) is crippled for life. How are we supposed to live? How is he (she) supposed to make a living?"

Human Resource Director: "That is what workers' compensation insurance is for. All medical bills will be paid, and we are sure the lump sum payment and/or vocational training will provide the skills needed to find another job."

You: "You know the workers' compensation payment will never repay this loss. He (she) will always be without a hand. I want to speak to the Safety Director. How could he or she allow this to happen?"

Company President: "On advice of our attorney, we cannot allow that. I know you are deeply troubled by this terrible accident, but it was an accident. You may wish to try legal action against the machine manufacturer, but since your _____ failed to replace the guard, I don't think there is much recourse. Again, we at the company are terribly sorry, but again, these things happen. Unfortunately, I must go to another meeting, but we wish you all the best."

Now, again, look at this injured employee as someone close to you! How do you feel about this situation? Was there something the company could have done? What about an OSHA citation? Is it enough, even though the company was appealing it? Let's start with what the company could have done.

## Corporate Responsibility and Accountability

Was there something that the company could have done to prevent this accident? Of course there was. The company knows it, OSHA knows it, and so do you. There is, however, very little the injured employee and relatives can do about it. He or she will receive full medical care and a lump sum payment for the loss of an extremity. He or she may also receive vocational training for a future job.

The company may also discharge the Safety Director for malfeasance. However, that will most certainly not occur until after the appeal hearing. Firing the Safety Director prior to that would give the impression that he or she—hence, the company—was at fault in this tragedy.

Why did OSHA issue a serious citation to this company? Was a serious violation sufficient for the accident? A willful violation must prove that the company willfully and knowingly put the employee at risk. In this case, the company did not do that. The accident was just a terrible oversight on the part of the maintenance department, and your relative.

Stop now and think about how you would feel. Is your loved one going to receive adequate compensation for this life-altering injury? How about the family and extended family? Will their lives be changed as well?

Please understand that this scenario is played out every day in the United States. Who was at fault in this case? Wasn't the maintenance department trained properly, or were they just in a hurry? What about the Safety Director—what is his or her role in all of this? Has he or she done an adequate job?

What if you were the Safety Director? Would this be a big deal for you? Or would you feel that it was nothing you could have prevented, and that you were in no way responsible? How would you feel about yourself after this occurred?

I recently had lunch with a Safety Director who told me of a fatality at one of his East Coast plants. A Line Supervisor at his company was unaware that an overhead crane was in operation. The crane operator did not see the Line Supervisor and the crane and its load of steel crushed the Supervisor's head. OSHA cited the company for not having a warning device on the crane. My friend spoke with the Safety Director regarding the incident and was appalled at his attitude. His position was that there was nothing he could have done to prevent the injury. The crane operator should have been more observant and yelled out a warning to the Line Supervisor. My friend told the Safety Director that he personally wouldn't be able to go back to work if he had allowed an employee to be killed. The Safety Director responded, "I didn't kill the guy, the crane did. I am not responsible for every injury that happens here. I refuse to feel guilty over this guy's death." My friend's parting comment was poignant: "I hope you were the one who had to tell his wife that he wasn't coming home."

Who do you agree with in this conversation? True, Safety Directors simply cannot prevent all accidents. But shouldn't Safety Directors accept some level of responsibility in preventing them? I would not want a surgeon operating on me with the same attitude exhibited by the Safety Director who refused to feel any guilt about the fatality. How you feel depends on your personal commitment to the protection of your employees.

There is a closing remark that attorneys often use in personal injury cases to send the message home to the jury. "God made this person the way he *was*. The company made him the way he *is*."

# Every Day in the USA

Let's spend a few pages looking at what is really happening in the American workplace regarding occupational safety.

As I mentioned previously in the book, an American worker is killed every 99 minutes. One is injured every 9 seconds.

It costs money for this many workers to be killed and injured:

- Wage and productivity losses = $59.8 billion/year

- Medical costs =$19.2 billion/year

- Administrative costs =$25.5 billion/year

**Lost Time Expense**

- 120,000,000 days lost/year

**Who We Are Killing**

- Almost 50 percent of all work-related deaths occur among workers between the ages of 25 and 44.

- Other than highway vehicle accidents and workplace violence, the number one killer in the workplace is being struck by an object, followed by falls to a lower level.

## Does the American Worker Really Need OSHA?

Let's spend some time looking at what is going on in American industry regarding workers' safety and health. Remember the three categories of business owners I came up with: "Catch Me If You Can," "Just Enough To Get By," and "Safety Smart."

Well, let's see what OSHA is doing about the "Catch Me If You Can" owner. I receive numerous newsletters and mailings giving updates on occupational safety engineering. They include citations that are being issued to American businesses and reasons for the citations.

A company in Ohio was reported to OSHA's attention via an employee complaint. Their total fines were more than $400,000. They had a total of 6 willful citations and 76 serious citations. I will only list the willful ones to save space:

- No employee training on energy control procedures.

- No written energy control procedures.

- No safe energy control practices.

- No electrical safe work practices.

- No hearing conservation program.

- No guards on equipment that could collapse and hit employees.

A New Jersey company that had seven amputations over a four-year period was fined more than $262,000. Their citations included 3 willful and 39 serious citations:

- No machine guards.

- No power press inspections.

- No hearing conservation program.

- Locked or blocked fire exits.

- No lockout/tagout program.

- Lack of personal protective equipment.

An Illinois company that was inspected by OSHA after being contacted by an employee was fined over $720,000 for numerous willful citations, including:

- Failure to report injuries to OSHA.

- Failure to report injuries to state workers' compensation system.

- No hearing conservation program.

- No machine guards.

- Locked exit doors.

- Failure to provide eye protection to employees.

These are three examples from a one-month's period of publications. Whether or not we need OSHA can best be summed up by this thought: How bad would it be if there were no regulatory agency to respond to employee complaints? How much support would the Safety Director receive if there were no regulatory consequence for not protecting employees? Would there even be a Safety Director or a safety program? Sometimes the best motivator is the fear of being caught!

## Making a Decision

Serving as your company's Safety Director is very different from any other job. Making a mistake or error in judgment in this role can cause an injury or cost a life. It has been my belief for many years that the reason an American worker is killed every 99 minutes is because some Safety Directors are not adequately doing their jobs. It isn't that we don't know how to protect our workers—we do. We have the technology and education available to us. In many cases, these high incident rates are because those assigned to this important task simply lack the care, concern, or commitment they need to protect employees. Many Safety Directors just aren't that involved. For whatever reason, they don't have the ethical courage to stand up for what they know to be right. They do not have the moral courage to resign from the position if they can't do the job properly. I wish this weren't true, but often it is. The victims, of course, are the employees who either don't know or don't care about the frustrations of your position. Maybe it is easier to lie back and go through the motions of the job. But is it morally right? You know and I know that it is not. Make your decision now about your commitment to the safety profession. *Yes, it is that big of a deal!*

Now, let's turn to the last chapter of this book for a summary.

# 15

# "Now, What Is It I'm Supposed to Do Again?"
# Putting It All Together

Let's now review, chapter-by-chapter, the key points addressed in the book.

## You and Your Boss

- Does your company management have commitment to the program?

  - How do you measure that commitment?

  - How has that commitment been demonstrated to you?

- What are the motivating factors in your company's desire to have a safety program? Insurance costs? Fear of OSHA citations? Corporate mandate?

- What category does your boss fall into?

  - Isn't informed, but cares.

- Isn't informed, and doesn't care.

- Is informed, and cares.

- Is informed, and doesn't care.

- Methods to get the attention of your boss:

    - Determine the costs of your workers' compensation insurance.

    - Determine the lack of compliance with OSHA standards.

    - Determine your company's Incident Rate compared to other similar businesses.

- The five questions you need to ask yourself:

    1. Why were you selected as the company Safety Director?

    2. What authority and responsibility were you given?

    3. What support were you promised?

    4. What do you want to accomplish?

    5. What does your boss really want from the safety program?

Your answers to these points will determine if your goals and objectives are attainable and realistic for your situation.

## The Plant Manager

- Meet with your boss and explain the support you need from the Plant Manager.

- Define and apply the Four Responsibilities of the Plant Manager in the safety program:

    1. To assist the Safety Director in the adoption and enforcement of company rules and regulations.

2. To oversee First Line Supervisors to ensure acceptance, adoption, and enforcement of the safety program by the production and engineering of new equipment.

3. To accompany the Safety Director on all production-related safety inspections.

4. To be a permanent member of the safety committee.

- Does the Plant Manager ensure his or her supervisors' compliance with their seven guidelines? (See Chapter #3)

- Are you using discernment in your daily interaction with the Plant Manager?

## First Line Supervisors

- Are Line Supervisors following the seven guidelines?

1. Do the Line Supervisors ensure their employees' compliance with safety rules and regulations?

2. Do the Line Supervisors condone unsafe behavior for the sake of production?

3. Are the Line Supervisors using their acquired safety knowledge?

4. Are the Line Supervisors listening to their employees' ideas on ways to improve safety in the department?

5. Do the Line Supervisors solve safety problems actively, resourcefully, and independently?

6. Do the Line Supervisors make their rounds?

7. Do the Line Supervisors properly handle accident/incident investigations and other safety-related paper work?

- Are the Line Supervisors meeting their objectives:

    - Responsibility?

    - Accountability?

    - Consequences?

    - Knowledge?

## Line Employees

- The Two Critical Rules to Follow:

    1. Talk straight, and play fair.

    2. Be firm, fair, and consistent.

- Start with walking the plant floor and training so your employees get to know you and you get to know them.

- Three kinds of Safety Directors:

    1. *Sailors*: They don't get wet from the day-to-day waves of their program. They are more concerned about protecting themselves from political consequences.

    2. *Surfers*: Surfers represent the majority of Safety Directors. They keep one eye on their program and the other on their job security. They balance their safety efforts against the political climate.

    3. *Swimmers*: These Safety Directors are totally committed. Their motto is, "Whatever it takes." They are only concerned about protecting their employees, regardless of the political risks.

- Learn your plant's operation. Walk the floor; determine what types of machines are being used. What are the exposures to the employee? If you don't know, *ask!*

- In Safety Committee meetings, only deal with safety-related issues. Human Resources issues should be addressed by that office.

- You cannot please all of your employees. Remember the -10, +10, and the ± 80. Your focus is always on the ± 80.

## Technical Knowledge

- Learn to research and interpret the OSHA Standards.

- Good resource and technical books to study:

  1. National Safety Council's *Accident Prevention Manual.* Two Volume Set.

  2. National Safety Council's *Fundamentals of Industrial Hygiene.*

  3. Complete *Manual of Industrial Safety.*

- No company, no product, no firm, no consultant is OSHA approved. None!

- Obtain certification for the right reason. Two organizations to consider joining:

  - The American Society of Safety Engineers (ASSE)

  - The World Safety Organization (WSO)

- Realistic certification is the needed experience and the necessary technical knowledge to properly do the job. "Stuff and Spew" seminars may get you ready for a test, but not for the real world.

- "Hub cap safety" makes your program look good, but it does very little to reduce injuries and protect workers.

## "Nobody Cares"

- What you do after you hit your ceiling and feel frustrated and overwhelmed will determine the future success of your program.

Will you decide you can do no more, or will you make a decision to turn yourself and your program up a few notches?

- Who are "They" who are not giving you the support you need?

- Sit down and make a list of who "They" are and what they are doing or not doing specifically that is blocking your program.

- When your boss says, "Do the Best You Can," document for him or her what it is you are already doing and the time it takes to do so. Document what other activities you need to do in order to provide the type of program that is needed.

- While documenting your activities, remember the "Boss's Dozen Responses."

    1. Oh, give me a break! Those times are grossly inflated.

    2. Safety doesn't take that much time.

    3. Look out—he (she) wants a raise.

    4. Why can't he pull his (her) own weight like everybody else?

    5. If he (she) thinks I'm going to hire another employee, he's (she's) nuts!

    6. Why can't he (she) just stop whining and do his (her) job?

    7. I wonder what's really bothering him (her)?

    8. I'll call one of my golfing buddies; his HR Director handles safety and doesn't whine about it.

    9. If he (she) wants the job, he (she) will just have to make do.

    10. What's the worst that can happen if things go along as they are?

    11. Who else can I get to replace this sniveler?

    12. I'll just tell him (her) to "Do the best you can."

- Keep in mind the "Ten Comments You Don't Want To Hear" from your boss:

  1. Why didn't you tell me?

  2. Were you aware of this? If not, why not?

  3. My door was always open to you.

  4. Didn't I tell you that I supported the safety program?

  5. What have you been doing all this time?

  6. If you needed more time, why didn't you say so?

  7. I thought you were responsible enough for this position.

  8. I am very disappointed in you.

  9. Couldn't you have done anything to prevent this?

  10. Under the circumstances, we are going to have to let you go.

- Remember the three Ps: *Be poised, prepared, and professional.*

## The Disciplinary Process

- As the Safety Director, you should be involved with the disciplinary process as an objective professional. Neither the activist/enforcer nor the hands-off approach will support your program.

- Beware of "Linkage Mentality." Employees who commit unsafe acts and injure themselves are not absolved of disciplinary action by nature of the resultant injuries.

- What policy and documentation does your company really have on safety rule violations?

- Develop a workable and pragmatic disciplinary system that is understood and implemented.

- What is the consequence for creating an unsafe condition or committing an unsafe act at your company?

# The Safety Committee

- A successful safety committee aids both management and the employees. If it doesn't serve both, it doesn't serve the safety program.

- The successful safety committee has:

  1. A meaningful purpose.

  2. Effective staffing and structure.

  3. Support while carrying out its mission.

  4. Active participation and cooperation.

- The responsibilities of the safety committee include:

  1. Monitoring tasks.

  2. Educational tasks.

  3. Investigative tasks.

  4. Evaluative tasks.

- Remember the Ten Steps to Running a Constructive Safety Committee:

  1. Develop a clear mission statement and stick to it.

  2. Plan the meeting carefully and write an agenda.

  3. Keep the meeting short and to the point.

  4. Keep the meeting respectful and friendly.

  5. Don't argue a point.

6. Be honest.

7. Always solicit suggestions.

8. Recap the meeting.

9. Make time to ask and answer questions.

10. Follow up on unanswered questions.

- The most important element prior to concluding a safety committee meeting is to determine *who does what, when?*

## Company Politics

- Determine what the problem is for you and your program.

- Even if the company "Oracle" is obstructing your efforts, documentation will support your position.

- Document action or inaction, event, consequence, and needed correction in an objective format.

- Use objective data to persuade family members to assist in controlling errant family members.

- Gender is not an issue regarding the qualifications of a Safety Director.

- Determine your company's "prevention philosophy." Are you pro-active or reactive?

## Workers' Compensation

- Elimination vs. Containment. You are not going to eliminate worker's fraud and abuse. You can, however, contain it.

- You and your company are the people in the "white hats." It is not your job or prerogative to deny a claim. It is the responsibility of your insurance carrier and the medical care provider to investigate possible fraudulent cases.

- Finger Tip Facts of Workers' Compensation:

    - An injury is work related if it occurs "in and out of the course of employment."

- The four types of disabilities for your employees are:

    1. Temporary partial: Usually the employee will reach MMI and return to work.

    2. Temporary total: The employee will be limited in some way, but will recover.

    3. Permanent partial: The employee will be limited permanently in some ability.

    4. Permanent total: The employee will not be able to return to work because of the injury.

- Exceptions to the Exclusive Remedy Rule:

    - Fellow employee suits: Employees can sue other employees who they allege caused them injury. (Check your state laws.)

    - Bringing suit against a contractor: A suit against another firm causing injury.

    - Bringing suit against equipment manufacturer: The most popular type of suit which is against the manufacturer of the equipment that caused an injury.

- Return-to-Work in the Real World:

    - Plant Managers: Remember the Line Supervisors and employees are their colleagues. Be sensitive to that. Explain the financial benefits to them.

- First Line Supervisors: Realistically, First Line Supervisors have nothing to gain in the RTW program. Emphasize the why, what, and who of the program.

- Line Employees: Remember, the "nothing is real unless it's personal" approach in dealing with employees. The RTW must be absolutely fair and consistent or it won't be accepted.

- Injured Employees: Never violate the doctor's restrictions on the employee. Treat injured employees with dignity and respect. Remember, they didn't deliberately injure themselves. Some way, somehow your safety program allowed them to be injured.

- Stay in touch with injured employees. How is the medical treatment? Are they getting their checks on time? Is the employee's family having any difficulty? Workers' Compensation Attorneys make millions doing what should be standard procedure for your company.

## Employee Training

- OSHA's Voluntary Guidelines:

  - Determine whether training can solve injury problems.

  - Determine what training is needed.

  - Ascertain goals and objectives.

  - Plan learning activities.

  - Manage training.

  - Determine productiveness of training.

  - Modify training based on feedback.

- How to Train:

  - Be wary of "quick fix" videos for a high price. They almost never do the job.

  - Remember, hold your supervisors and your employees personally accountable for learning the safety subject you are teaching. Use learning exercises for this purpose.

  - Use of training handouts that require participants to fill in the blank are an excellent and inexpensive way to train. They also create responsibly and accountability on the part of participants.

  - Supervisors should always attend their employees' training sessions. They are responsible for their employees, not you!

## Accident Investigation

- The six basic questions of accident investigation are:

  Who, what, when, where, how, and why.

- Supervisors must be personally involved in the accident investigation process. It is your job and that of the Plant Manager to ensure they do the accident investigation properly.

  - You are fact finding, not fault finding.

  - Sit one-on-one with the employee:

    1. Ask about medical care.

    2. Ask for help in the investigation.

    3. Ask open, not accusatory questions.

  - The Line Supervisor and you must demonstrate positive facial affect and body language to the employee.

–   Determine how the safety system failed.

–   Solicit questions, comments, observations or suggestions to help the investigation.

–   Ask the employee, "What would you have done differently?"

–   Get the word out to the rest of the company how the injury occurred and what is being done to prevent recurrence.

# Communication

-   Safety Communication Obstacles:

    –   Production noise.

    –   Language.

    –   Planning communication.

-   Organize your communication:

    –   Make requirements understandable.

    –   Remember your audience.

    –   Be professional.

    –   Know when to say what.

    –   Be the expert on your safety policies and procedures.

-   You can't hear when you talk.

    –   Be a good faith listener.

    –   Listen to what the other person is saying.

    –   Don't jump to conclusions.

    –   Know your facts.

- Do you understand what's being said?

  - Get feedback to ensure comprehension.

  - Ask questions.

  - Give praise.

- Planning a safety meeting:

  - Determine why you are having the meeting.

  - Determine who needs to attend your meeting.

  - Determine how much time you need

- Running the meeting:

  - State the purpose of the meeting immediately.

  - Prime the pump by giving attendees background on the issue.

  - Blend and harmonize the comments of participants.

  - Summarize the meeting.

  - Remember to conclude with who does what, when.

- Dress for work:

  - Safety Directors aren't Wall Street brokers; some just think they are. Dress down.

  - The three E's for dress: *employees, equipment, and exposures.*

## Personal Commitment

- There are two common denominators of ineffective Safety Directors:

1. They are more worried about protecting their position than with employees' safety and lives.

2. They don't know what they don't know.

- Making your decision: Determine if this is the position you want, and what you are prepared to commit to it. Remember, human lives are at stake.


I titled this book *Safety Is A People Business* so the reader would immediately know that the safety profession is, indeed, about people. It is about the people you work with, the people you are paid to protect, and you. As I stated earlier, safety is an engineering discipline, but it is also a behavioral science.

I have never seen a successful safety program in which the Safety Director was some mad monk engineer sitting in his office pouring over incident rates. Safety Engineering is about getting out on the plant floor—looking, talking, listening, and relating to people. If your employees see you as talking straight and playing fair and showing concern and commitment to them, your chances of success are greatly improved. You cannot reach that level of success without being on the floor with them. Only then can they see and share your commitment to safety.

I extend to you my sincere best wishes for that success.

# Appendix A

**Answers to Standard Questions in Chapter 5.**

1. 1910.101 (a)

2. 1910.157 (d)(3)

3. 1910.178 (n)(1)

4. 1910.212 (a)(3)(i)

5. 1910.215 (a)(4)

6. 1910.1020 (e)(1)(i)

7. 1910.25 (c)(2)

8. 1910.133 (a)(1), (a)(2)

9. 1910.134 (b)(1), (e)(3)

10. 1910.66 (g)(3)(ii)

# Glossary

| | |
|---|---|
| **Accident** | An unplanned event, not necessarily resulting in injuries or damage to property, which interrupts the activity in process. |
| **ASSE** | American Society of Safety Engineers. |
| **Attentiveness** | Concentrating one's attention on something. |
| **CFR** | Code of Federal Regulations<br>29 CFR 1910     General Industry<br>29 CFR 1926     Construction |
| **Claims representative** | The contact from the insurance carrier who handles claims services for treatment, rehabilitation, and workers' compensation benefits for the injured employee. |
| **Commitment** | Something one is bound to do or forbear. |
| **Communication** | Something conveyed by writing, speech, or signals. |
| **Compliance officer** | A state or federal occupational safety and health official. |
| **Comportment** | One's actions in general or on a particular occasion. |
| **Discernment** | Acuteness of perception or judgment. |
| **3 E's for dress** | Employees, equipment, and exposures. |
| **Egress** | A place or means of going out. |

| | |
|---|---|
| **Exclusive remedy** | The exclusive recourse of the employee to eliminate legal action by the employee against the employer. |
| **Experience Modification Factor** | A device used to measure a company against other like businesses with the same type of exposure by using a premium financial factor. |
| **General Duty Clause** | Each employer shall furnish its employees a place of employment that is free from recognized hazards that are causing or likely to cause death or serious physical harm to the employees. |
| **Incident** | An undesired event that may cause personal harm or other damage. |
| **Incident Rate** | A measuring device for occupational injury and illness occurrences that companies can use to compare themselves against like companies. |
| **Industrial Hygiene** | The science devoted to the recognition, evaluation, and control of those environmental factors or stresses that may cause sickness, impaired health, or significant discomfort to employees or residents of the community. |
| **Insurance Agent** | A person who serves as the bridge between the company and the insurance carrier. |
| **MMI** | Maximum Medical Improvement. |
| **OBW** | "Oh, By the Way" you are responsible for safety. |
| **OSHA** | Occupational Safety and Health Administration. |
| **OSHA Standards** | Rules and regulations that apply to safety. |
| **Point-of-operation** | The area on a machine where work is actually performed upon the material being processed. |

| | |
|---|---|
| **R.A.C.K.** | Responsibility, Accountability, Consequences, and Knowledge. |
| **Recognized hazard** | A condition that is of common knowledge or general recognition in the particular industry in which it occurs and is detectable by means of senses, or such wide, general recognition as a hazard in the industry that there are generally known and accepted tests for its existence that should make its presence known to the employer. |
| **Return-to-work (RTW)** | Policies and procedures used by a company to return an employee to work while the employee is recovering from an occupational injury or illness. |
| **SIC** | Standard Industrial Classification code. |
| **200 Log** | A required form from the Occupational Safety and Health Administration to record occupational injures and illnesses. |
| **Willful violation** | When OSHA determines that the employee knew there was a hazardous condition that could kill or cause serious injury and didn't correct it. |
| **Workers' Compensation** | Insurance coverage for employees in the event they are injured and/or become ill in and out of the course of employment. |
| **WSO** | World Safety Organization. |

# Bibliography

Bureau of Business Practice. *The BBP Safety Management Handbook.* Waterford, CA: Prentice Hall, 1990.

Davis, Weldon, Grubbs, John R., and Nelson, Sean M. *Safety Made Easy, A Checklist Approach to OSHA Compliance.* Rockville, MD: Government Institutes, Inc., 1995.

Dennis, Leslie E. and Onion, Meredith L. *Out In Front: Effective Supervision In the Workplace.* Itasca, IL, National Safety Council, 1990.

J.J. Keller & Associates, Inc. *Compliance Audits: Essential Checklists for OSHA, EPA and Other Key Agencies.* Neenah, WI: J.J. Keller & Associates, Inc., 1994.

Kohn, J.P., Friend, M.A., and Winterberger, C.A., *Fundamentals of Occupational Safety and Health.* Rockville, MD: Government Institutes, Inc., 1996.

Manning, Michael. *"So You're the Safety Director!" An Introduction to Loss Control and Safety Management.* Rockville, MD: Government Institutes, Inc., 1996.

Mansdorf, S.Z. *Complete Manual of Industrial Safety.* Englewood Cliffs, NJ: Prentice Hall, 1993.

The Merritt Company. *OSHA Reference Manual, Occupational Safety and Health Compliance Simplified.* Santa Monica, CA: The Merritt Company, 1993.

National Safety Council. *Accident Facts*, 1996 Edition. Itasca, IL: National Safety Council, 1996.

-- Accident Prevention Manual for Business & Industry: Administration & Programs, 11 Edition. Itasca, IL, NSC, 1996.

-- Accident Prevention Manual for Business & Industry: Engineering & Technology, 11 Edition. Itasca, IL, NSC, 1996.

-- *Protecting Workers' Lives, 2nd Edition: A Safety and Health Guide for Union.* Itasca, IL, NSC, 1992.

-- *Supervisors' Safety Manual, 8th Edition.* Itasca, IL: NSC, 1993.

Pierce, David F. *Total Quality for Safety and Health Professionals.* Rockville, MD: Government Institutes, Inc., 1995.

Plog, Barbara A., Niland, Jill, Qinlan, and Patricia J. *Fundamentals of Industrial Hygiene.* Itasca, IL: National Safety Council, 1996.

Powell, Colin. *My American Journey.* New York, NY: Random House, 1995.

Reich, Robert B. *Locked in the Cabinet.* New York, NY: Alfred A. Knopf, Inc., 1997.

Walsh, James. *Workers' Comp for Employers; How to Cut Claims, Reduce Premiums, and Stay Out of Trouble.* Santa Monica, CA: Merritt Publishing, 1993.

# Index

"One of the most dangerous things you can do is show up for work!"

– Dr. Michael V. Manning

"Safety is not an event, it's a *process*."

– Michael Melnik

"Nothing changes, if nothing changes."

– Ernie Larsen

| PC # | ENVIRONMENTAL TITLES | Pub Date | Price |
|------|----------------------|----------|-------|
| 585 | Book of Lists for Regulated Hazardous Substances, 8th Edition | 1997 | $79 |
| 4088 | CFR Chemical Lists on CD ROM, 1997 Edition | 1997 | $125 |
| 4089 | Chemical Data for Workplace Sampling & Analysis, Single User | 1997 | $125 |
| 512 | Clean Water Handbook, 2nd Edition | 1996 | $89 |
| 581 | EH&S Auditing Made Easy | 1997 | $79 |
| 587 | E H & S CFR Training Requirements, 3rd Edition | 1997 | $89 |
| 4082 | EMMI-Envl Monitoring Methods Index for Windows-Network | 1997 | $537 |
| 4082 | EMMI-Envl Monitoring Methods Index for Windows-Single User | 1997 | $179 |
| 525 | Environmental Audits, 7th Edition | 1996 | $79 |
| 548 | Environmental Engineering and Science: An Introduction | 1997 | $79 |
| 578 | Environmental Guide to the Internet, 3rd Edition | 1997 | $59 |
| 560 | Environmental Law Handbook, 14th Edition | 1997 | $79 |
| 353 | Environmental Regulatory Glossary, 6th Edition | 1993 | $79 |
| 562 | Environmental Statutes, 1997 Edition | 1997 | $69 |
| 562 | Environmental Statutes Book/Disk Package, 1997 Edition | 1997 | $204 |
| 4060 | Environmental Statutes on Disk for Windows-Network | 1997 | $405 |
| 4060 | Environmental Statutes on Disk for Windows-Single User | 1997 | $135 |
| 570 | Environmentalism at the Crossroads | 1995 | $39 |
| 536 | ESAs Made Easy | 1996 | $59 |
| 515 | Industrial Environmental Management: A Practical Approach | 1996 | $79 |
| 4078 | IRIS Database-Network | 1997 | $1,485 |
| 4078 | IRIS Database-Single User | 1997 | $495 |
| 510 | ISO 14000: Understanding Environmental Standards | 1996 | $69 |
| 551 | ISO 14001: An Executive Repoert | 1996 | $55 |
| 518 | Lead Regulation Handbook | 1996 | $79 |
| 478 | Principles of EH&S Management | 1995 | $69 |
| 554 | Property Rights: Understanding Government Takings | 1997 | $79 |
| 582 | Recycling & Waste Mgmt Guide to the Internet | 1997 | $49 |
| 594 | Texas Environmental Regulations Manual | 1997 | $125 |
| 566 | TSCA Handbook, 3rd Edition | 1997 | $95 |
| 534 | Wetland Mitigation: Mitigation Banking and Other Strategies | 1997 | $75 |

| PC # | SAFETY AND HEALTH TITLES | Pub Date | Price |
|------|--------------------------|----------|-------|
| 547 | Construction Safety Handbook | 1996 | $79 |
| 553 | Cumulative Trauma Disorders | 1997 | $59 |
| 559 | Forklift Safety | 1997 | $65 |
| 539 | Fundamentals of Occupational Safety & Health | 1996 | $49 |
| 535 | Making Sense of OSHA Compliance | 1997 | $59 |
| 563 | Managing Change for Safety and Health Professionals | 1997 | $59 |
| 589 | Managing Fatigue in Transportation, *ATA Conference* | 1997 | $75 |
| 4086 | OSHA Technical Manual, Electronic Edition | 1997 | $99 |
| 598 | Project Mgmt for E H & S Professionals | 1997 | $59 |
| 552 | Safety & Health in Agriculture, Forestry and Fisheries | 1997 | $125 |
| 523 | Safety & Health on the Internet | 1996 | $39 |
| 597 | Safety Is A People Business | 1997 | $49 |
| 463 | Safety Made Easy | 1995 | $49 |
| 590 | Your Company Safety and Health Manual | 1997 | $79 |

◫ = Electronic Product available on CD-ROM or Floppy Disk

GOVERNMENT
INSTITUTES

PUBLICATIONS CATALOG
1997

## PLEASE CALL OUR PUBLISHING DEPARTMENT AT (301) 921-2355 FOR A FREE PUBLICATIONS CATALOG.

### Government Institutes
4 Research Place, Suite 200 • Rockville, MD 20850-3226
Tel. (301) 921-2355 • FAX (301) 921-0373
E mail: giinfo@govinst.com • Internet: http://www.govinst.com

# GOVERNMENT INSTITUTES ORDER FORM

*4 Research Place, Suite 200 • Rockville, MD 20850-3226 • Tel (301) 921-2355 • Fax (301) 921-0373*
Internet: *http://www.govinst.com* • E-mail: *giinfo@govinst.com*

## 3 EASY WAYS TO ORDER

**1. Phone:** **(301) 921-2355**
Have your credit card ready when you call.

**2. Fax:** **(301) 921-0373**
Fax this completed order form with your company
purchase order or credit card information.

**3. Mail:** **Government Institutes**
4 Research Place, Suite 200
Rockville, MD 20850-3226
USA
Mail this completed order form with a check, company
purchase order, or credit card information.

## PAYMENT OPTIONS

❑ **Check** (*payable to Government Institutes in US dollars*)

❑ **Purchase Order** (this order form must be attached to your
company P.O. Note: All International orders must be pre-paid.)

❑ **Credit Card** ❑   ❑   ❑

Exp.____/____

Credit Card No. _____

Signature _____
Government Institutes' Federal I.D.# is 52-0994196

## CUSTOMER INFORMATION

**Ship To:** (Please attach your Purchase Order)

Name: _____

GI Account# (*7 digits on mailing label*): _____

Company/Institution: _____

Address: _____
(please supply street address for UPS shipping)
_____

City: _____ State/Province: _____

Zip/Postal Code: _____ Country: _____

Tel: (____) _____

Fax: (____) _____

E-mail Address: _____

**Bill To:** (if different than ship to address)

Name: _____

Title/Position: _____

Company/Institution: _____

Address: _____
(please supply street address for UPS shipping)
_____

City: _____ State/Province: _____

Zip/Postal Code: _____ Country: _____

Tel: (____) _____

Fax: (____) _____

E-mail Address: _____

| Qty. | Product Code | Title | Price |
|------|--------------|-------|-------|
|      |              |       |       |
|      |              |       |       |
|      |              |       |       |
|      |              |       |       |
|      |              |       |       |
|      |              |       |       |

❑ **New Edition No Obligation Standing Order Program**
Please enroll me in this program for the products I have ordered. Government
Institutes will notify me of new editions by sending me an invoice. I understand
that there is no obligation to purchase the product. This invoice is simply my
reminder that a new edition has been released.

**15 DAY MONEY-BACK GUARANTEE**
If you're not completely satisfied with any product, return it undamaged
within 15 days for a full and immediate refund on the price of the product.

Subtotal_____
MD Residents add 5% Sales Tax_____
Shipping and Handling (see box below)_____
**Total Payment Enclosed**_____

| **Within U.S:** | **Outside U.S:** |
|---|---|
| 1-4 products: $6/product | Add $15 for each item (Airmail) |
| 5 or more: $3/product | Add $10 for each item (Surface) |

**SOURCE CODE: BP01**